# MBA Math & More

## Concepts You Need in Your First Year of Business School

CHRIS RYAN AND CARRIE SHUCHART

10 9 8 7 6 5 4 3 2 1

ISBN: 978-1-5062-4753-3

# Table of Contents

# PREFACE

You may already be in your first year of business school. Or maybe you haven't started school yet.

Either way, we were once where you are. We remember what it was like.

Back before we started our respective MBA programs, neither of us had spent much time thinking about the *actual* future. What were we really going to *do* in business school?

We both came from nontraditional backgrounds, and our images of business school were a bit fuzzy. We knew there'd be classes and clubs, job hunts and jaunts to exotic locals, but what were we actually going to do?

The reality turned out to be much more complicated. Countless imperatives to juggle. A zillion different directions to be pulled in. It was both glorious and maddening.

For us, and for so many folks we've talked to over the years, business school was truly transformative.

Here's the single bit of life advice we'll offer:

To get the most out of your experience in business school, try—and keep trying—to *let* it transform you.

Engage. Stretch yourself. Challenge the fixed assumptions you have about the way you are.

"Yes, yes," you say. "But I'm worried about the big accounting test."

Yes indeed. Business school is actually school. And in that capacity, it can be tough. Occasionally, even excruciating. Don't let anyone tell you differently.

The math-based subjects you encounter in your first year are absolutely no joke.

Even if you spent gobs of time with numbers before business school, you probably didn't do very many one-tailed tests of null hypotheses. (Don't worry—those are covered in Chapter 3: Statistics.)

Moreover, you need to tackle these academic challenges rapidly, in the midst of all sorts of other pressures.

For instance, you may be changing careers, so you need to network right away like crazy and prove your commitment to your new path, while you search for the right summer job . . . or just *a* summer job.

And although business schools tend to be responsive to students, they have the same structural flaw as most universities and colleges: professors are typically more rewarded for research than for teaching. That Ph.D. teaching you statistics may be a brilliant statistician. Unfortunately, he may not be a brilliant *teacher*.

In fact, the more ninja-level-stats chops he has, the more he may forget what it's like not to have any stats chops at all. Some call it the "expert trap": experts can't remember not being experts. So just as you're striving to master those one-tailed hypothesis tests, you may be struggling to get good help.

That's where this book comes in.

We're both teachers. Even when our job descriptions have said other things, teaching is what we love.

During business school, we spent many hours explaining material to our classmates. After business school, we taught umpteen GMAT and

GRE classes to aspiring grad students. And after that, they kept asking for help—which is why we wrote this book.

Our goal here is simple: to break down the most important concepts you'll meet in your first year of b-school—particularly the quantitative ones—so that you can settle in and truly be transformed by the experience.

Hence, the title *MBA Math & More.*

Let's get to it. Good luck!

*Chris & Carrie*

# MBA Math & More

# Grasping First-Year Academics

During orientation, you might at times forget that business school is actually school. Once classes start, this illusion will evaporate like the dew. Whether you're gunning for the top of your class or just looking to make it through, you've always got to deal with the *school* business of business school.

## A COMPLEMENT, NOT A SUBSTITUTE

These chapters are not meant to replace any course. You don't need an "MBA in a box"; after all, you're getting a real MBA. Rather, these chapters are intended to introduce the ***quantitative concepts involved in typical introductory courses.***

Our focus is on "quant" because that's where we can do the most good in the fewest pages. Even if you have a serious math background, we think you'll find this material useful.

Our focus is also on concepts because you'll get plenty of applied practice in your classes. The upcoming chapters attempt to paint the big picture, clarify the lingo, reveal assumptions, and point out traps.

## CHECKLIST FOR ACADEMIC PREP

| Chapter | Topics |
|---|---|
| 1: Excel & PowerPoint | ☐ Excel<br>☐ PowerPoint |
| 2: Economics & Game Theory | ☐ Microeconomics: Supply, Demand, & Elasticity<br>☐ Micro: Cost Curves<br>☐ Micro: Profit Maximization<br>☐ Macroeconomics: Circular Flow & Production Function<br>☐ Macro: Money Supply & Money Demand<br>☐ Macro: Investment, Savings, & General Equilibrium<br>☐ Game Theory |
| 3: Statistics | ☐ Descriptive Statistics<br>☐ Probability<br>☐ Distributions<br>☐ Sampling & Hypothesis Testing<br>☐ Correlation & Regression |
| 4: Accounting | ☐ The Big Three Financial Statements<br>☐ Transactions<br>☐ Managerial Accounting |
| 5: Finance | ☐ Time Value of Money<br>☐ Net Present Value (NPV) & Discounted Cash Flows (DCF)<br>☐ Capital Asset Pricing Model |
| 6: Marketing | ☐ Big Picture & Quant Tools |
| 7: Operations & Supply Chain | ☐ Process Analysis<br>☐ Queueing<br>☐ Inventory Management |
| 8: Common Threads | ☐ Common Threads & Key Tensions |

## WHAT'S LEFT OUT

Lots of interesting, valuable stuff is minimized or omitted, because you'll learn it just fine in school. For example, here's what you won't see much of in this book:

- Social psychology, organizational behavior, behavioral economics & finance, negotiations
  - E.g., systematic cognitive and emotional biases that affect decision-making
- Strategy & management theory
  - E.g., Porter's Five Forces and generic strategies
- The history of various companies and industries
- Leadership

Many of these subjects require hands-on application (e.g., negotiations and leadership). Moreover, you'll pick up bits and pieces of "business history" with every case that you do.

This isn't at all to say that these topics are less important than those listed on the opposite page. Rather, they don't lend themselves as well to guidebook treatment (and this book is already thick enough).

As you get into recruiting season, depending on your goals and the structure of the first-year curriculum at your school, you may need to accelerate your learning in some of the areas listed above. For instance, general strategy courses are sometimes placed as "capstones" at the end of the first year, but you should know about Porter's Five Forces long before any consulting interview. (By the way, these mysterious Five Forces are very easy to learn.) If necessary, borrow last year's strategy syllabus from a second year and review the highlights. Moreover, the consulting club at your school will almost certainly be aware of any sequence issues such as this one and will be able to help you out.

## HOW TO USE THIS PART

*Before school begins*: Only glide through these chapters. Do *not* try to master them. Just get the big picture. Skip around as you like.

*Once school begins*: Reread the material relevant to courses you're currently taking. It's that simple.

## THE BIG PICTURE OF BUSINESS

If you're like we were, you might not have much of a business background before school. Or you might have deep expertise in a specific industry or two. Either way, you're smart, but there is at least some knowledge of the business world that you lack. (Otherwise, why go back to school? Go conquer the world, since you already know everything.)

Your strategy and general management courses will seek to give you the "big picture of business." However, we want to get you started with a slightly different take on that picture.

## MULTIPLE LENSES

Bear with us, as a brief digression into physics is in order (when is it ever not?). *Wave-particle duality* is the idea that you can regard visible light either as a stream of particles[1] or as a rippling wave of electromagnetic fields.

Particle Model of Light

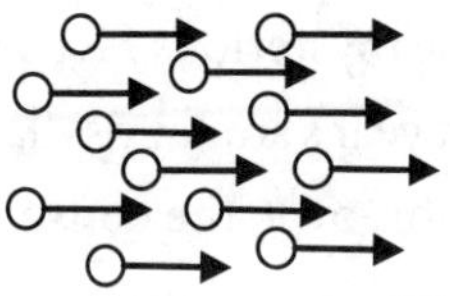

Wave Model of Light

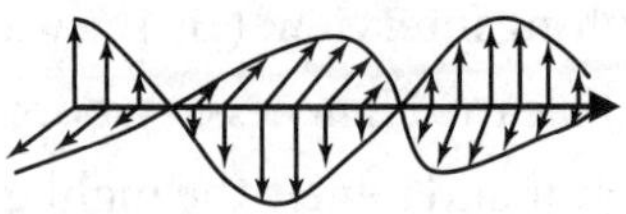

1 These particles are called photons. Science!

Which concept of light is right? Both and neither. They're competing metaphors that capture important aspects of the underlying phenomenon, but neither model is complete on its own. In fact, these models are complementary. When one is too limiting, the other works well.

Likewise, when you are a general decision-maker, you need to be able to apply ***multiple lenses*** to an issue. This "general management perspective" lies at the heart of the MBA degree. For instance, if you are deciding whether to launch a joint venture with another company, you need to consider many aspects of the venture—financial, operational, marketing, and so on—by gathering and synthesizing a variety of specialist opinions and analyses. No single analysis would be enough for this complex and weighty situation.

In the same vein, there are three simple models of that one basic thing you study in business school: a ***company***. Frankly, none of these models is that interesting by itself. Each is almost so simple as to be silly. That's why they're worth making explicit as a trio. The whole key is to have all three in your hip pocket and be ready to switch among them, like a particle physicist.[2]

## WHAT IS A COMPANY?

Just remember ***EFG***: ***Entity-Factory-Group***.

A company is...

1. A Legal ***Entity*** or "person"
2. A ***Factory*** or machine
3. A ***Group*** of people

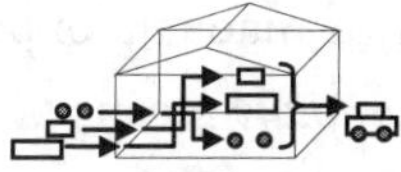

2 Presumably a better-dressed particle physicist.

1. A Legal ***Entity*** or "person"

The law considers a corporation an actual, living person in many respects. The root of the word *corporation* is *corp*, meaning "body." When you in*corp*orate your company, you turn it into a single *body* from the law's point of view.

But this is not the only reason to look at a company this way. When a business does something, you typically use a singular verb: *Coke* ***is*** *entering three new markets this year*. Of course, you don't picture some giant Coke can with feet storming into three new markets. But in analyzing the strategic actions of some company from the outside, you often need to simplify matters by thinking of the company as a monolithic actor with a single brain. (Yes, some companies seem to act as if there were only one brain on the premises, but that's a different matter.)

2. A ***Factory*** or machine

Peeling back the layers, you can also picture a company as an operating factory, which is itself really just a big machine with walls and a roof. Processes involving all kinds of stuff happen within the factory, which takes in raw materials and turns them into products for its customers. Money flows around as well: customers pay the company, which turns around and pays its suppliers.

The factory model can represent a literal factory making cars or a conceptual factory making ideas (e.g., a software company). Either way, in this functional view, you examine the way all the parts interlock.

3. A ***Group*** of people

Finally, of course, a company is made up of people. Back on the etymology kick, the word *company* means "a group of folks sharing bread." This is not to be touchy-feely. In a way, there is very little to any firm *but* its people. Every company decision, every company agreement, every company success or failure is ultimately the responsibility of some*one*, not some*thing*. You might even say that a company is a kind of shared idea, a web of expectations and promises among huge numbers of people (both those within the company and those on the outside). At all times, a company carries with it all the possibility and immense complexity of human interaction.

You can answer two other key questions about a company according to which of the three EFG models you're using.

**What does a company do?**

1. *Entity*: It plays a game in a complex ecosystem.
2. *Factory*: It produces products, selling them to customers.
3. *Group*: Nothing. It's the people, individually and collectively, who do everything.

**Why does a company exist?**

1. *Entity*: To survive and grow.
2. *Factory*: To produce and sell products at a profit.
3. *Group*: To fulfill various roles in society (a larger group of people) and meet people's needs.

You might naturally gravitate to one of these models more than the other two. That's fine; it tells you something about yourself.

You'll also find that your classmates take different positions. Some very existential arguments you hear in class boil down to clashing perspectives about what a company is, what it does, and what its purpose is.

These arguments get old very fast, as if you heard physicists arguing this way:

> Physicist #1: "Light's really a wave!"
> Physicist #2: "No, dummy! It's a bunch of particles!"
> *Repeat.*

Now replace the players with your classmates.

> Banker: "This company's just a legal entity."
> Engineer: "No, dummy! It's a bunch of interlocking parts!"
> HR Specialist: "Look, morons, it's all about the people."
> *Repeat.*

Ideally, when your classmates lock themselves into just one underlying model, you can recognize the stalemate and sail above the fray.

## COMMON THREADS—LATER ON

Chapter 8, at the end of this book, will draw together themes from across the next several chapters. Some important ideas will rear their beautiful heads more than once, as you'll see.

# Chapter 1: Excel & PowerPoint

This chapter provides an introduction to two mission-critical Microsoft applications: Excel and PowerPoint. While you're reading, pop open each application on your laptop so you can play along.

## EXCEL = MACHINE SHOP

You might think of Excel as a glorified calculator or as a table-maker.[1]

| Period | Sales (in millions) |
|---|---|
| Quarter 1 | $270 |
| Quarter 2 | $230 |
| Quarter 3 | $260 |
| Quarter 4 | $240 |
| Full Year | $1,000 |

Nice → table! (pointing to Quarter 2)

← Nice calculation! (pointing to Full Year $1,000)

That's all valid, but you should also think of Excel as a ***machine shop,*** where you build machines. (Stay with us on this for a few minutes—the analogy will help.) These machines act like little factories, taking in raw materials and spitting out finished products.

1 The program has all kinds of uses. Excel might even land you on the reality TV show "America's Next Top Modeler" (thanks, Follies).

**The Excel Machine Shop**

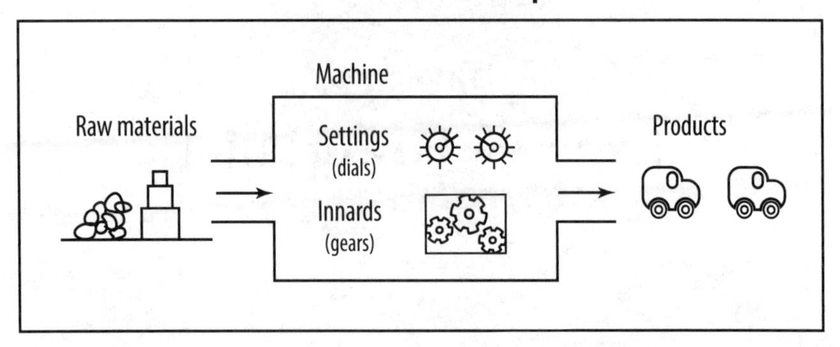


These machines are known as ***models***. Models typically have four components:

1. Raw materials for the machine = ***inputs*** for the model.
   These are the numbers you start with, also known as the data.
2. Settings on the machine = ***assumptions*** for the model. These are *also* numbers you start with, but ones you will change less often. A typical example is a general interest rate. Sometimes, the distinction between inputs and assumptions doesn't matter at all. Other times, it can be conceptually helpful to make this distinction.
3. Innards of the machine = ***intermediate calculations*** of the model. These are formulas that transform your raw numbers according to your settings. By definition, the innards are in the middle of the machine, so if the machine's working, you're just going to *watch* them as they spin around. Wheee!
4. Products of the machine = ***outputs*** of the model.
   These are the final calculated numbers you care about, the results of your last formulas.

Say you wanted to build a super-simple model that would tell you how much money you'd have in two years if you invested it at 5% interest, compounded annually. One version of the model has a single input and a single output.

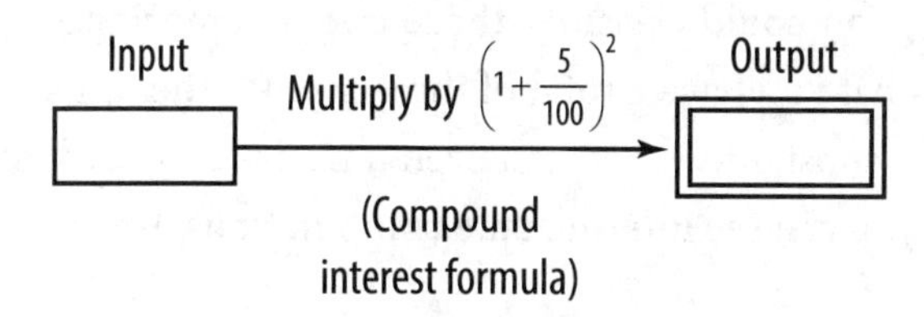


You plug in a dollar amount as the input, and you get the output.

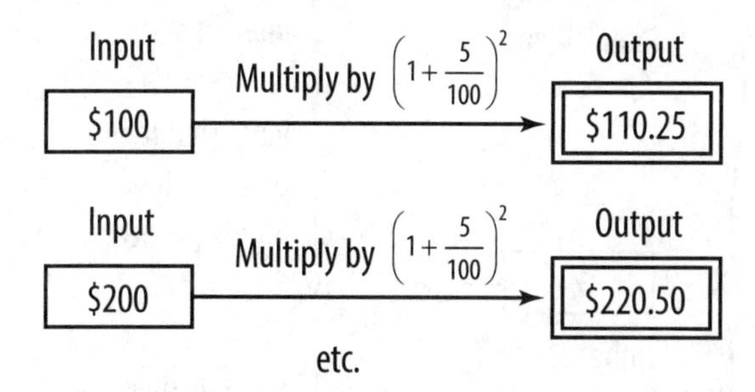


This model is equivalent to a ***function*** $y = f(x)$, where $y$ is the output and $x$ is the input. Specifically, this function $f(x)$ would be equal to $x \cdot \left(1+\frac{5}{100}\right)^2$; in other words, you multiply the input $x$ by $\left(1+\frac{5}{100}\right)^2$.

This model is fine, but not very flexible. What if you want to change the interest rate or the number of years? You should make them into assumptions that you can adjust:

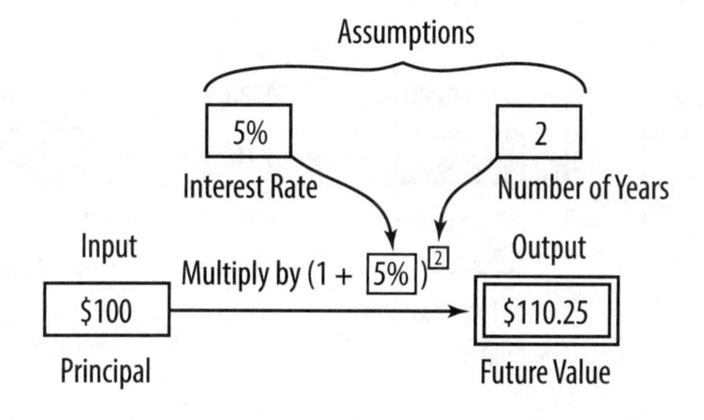

Optionally, you could consider those two assumptions to be inputs instead, so that you have a total of three inputs. The difference is only stylistic, not substantive. You can even introduce "innard" steps in the middle, so you can see intermediate computations. It's up to you.

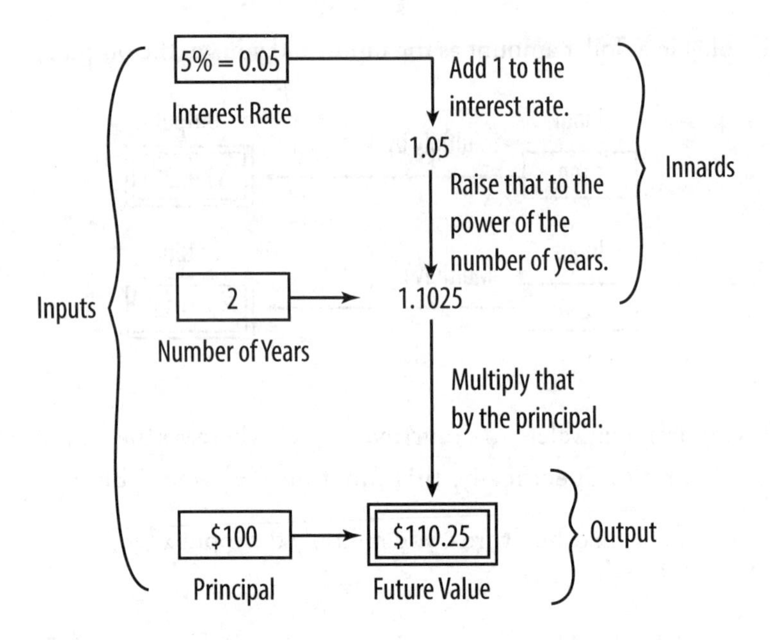


The key to making good models in Excel is ***deciding on the four components*** and then keeping them very straight. For instance, color-coding is amazingly helpful (assuming you're not color-blind).

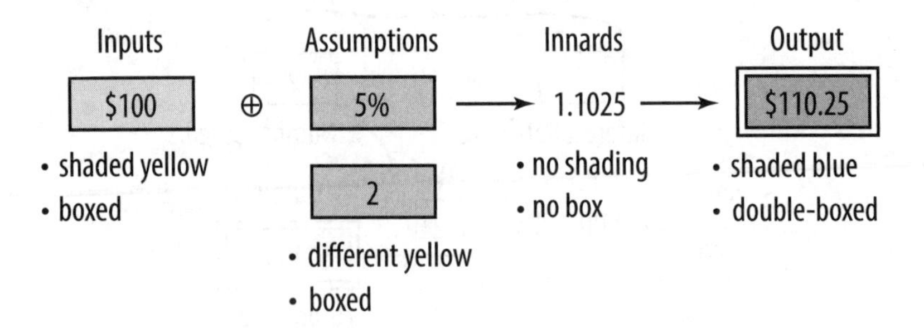


One of us got this specific shading and boxing advice over a decade ago. He can still easily decipher models he built back then, because the path from input to output is completely clear.[2] Models can be built in very different ways to accomplish the same goals, as shown by the simple examples above. Spend time up front to figure out what you want the inputs, assumptions, innards, and outputs to be, and you'll build the model faster and better. Plus, later on you'll be able to make adjustments easily, because you can figure out what you did.

## THE STRUCTURE OF EXCEL

When you open a new Excel file, you'll see a rectangular grid with lots of little boxes.

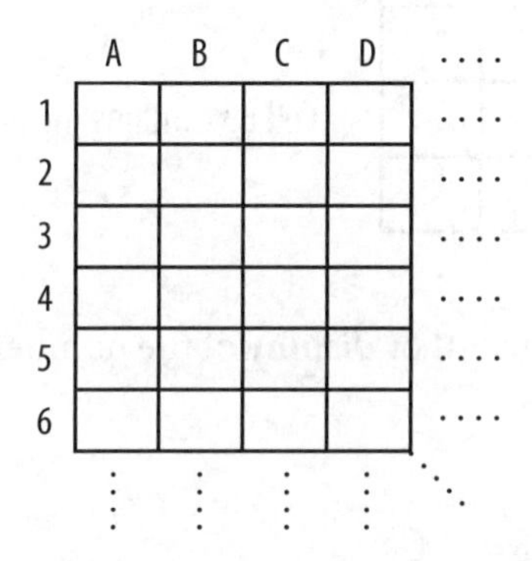


Each little box is called a ***cell***.

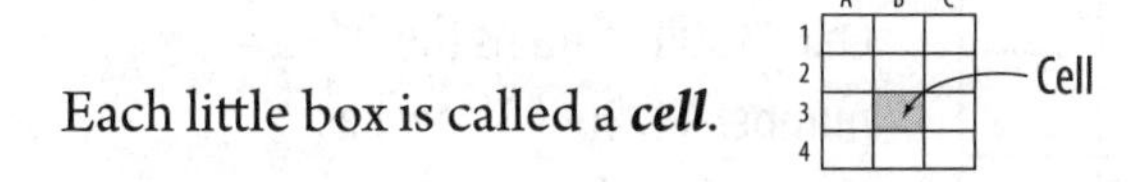


Cells are arranged in ***rows*** and in ***columns***.

Each cell is named by its column and row.

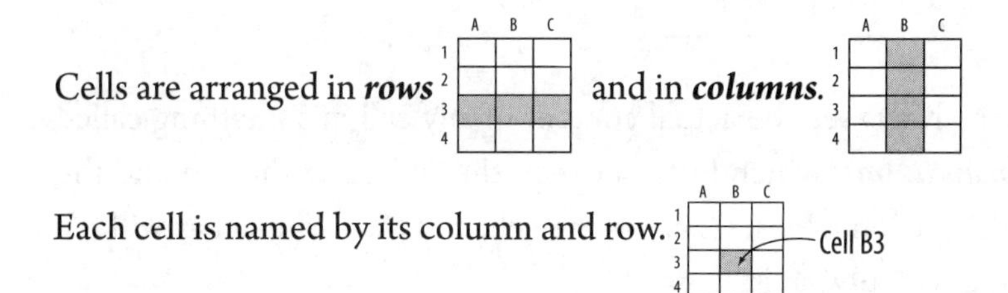


2 His Friday-night examinations of decades-old Excel models for proper shading and boxing reveal a thrilling social life.

Remember Battleship? When you want to hit a target, you announce the coordinates ("B6"). Cells are just like Battleship coordinates.

The whole grid is called a ***worksheet*** or a ***spreadsheet***, and you can have several worksheets in a ***workbook***.

A cell can contain one of three things:

1. A cell can contain a ***number***, such as an input or an assumption.

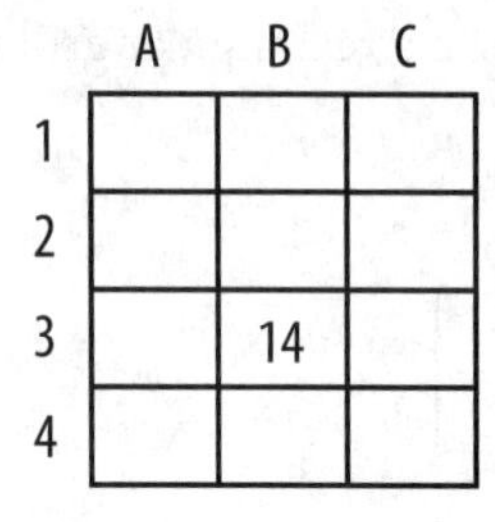


Cell B3 contains the number 14.

You can change the ***format*** or ***display*** of the number without changing its ***value***.

| | A | B | C |
|---|---|---|---|
| 1 | | | |
| 2 | | | |
| 3 | | $14.00 | |
| 4 | | | |

Cell B3 still contains the number 14, but B3 now has the currency format, so you see $14.00 in the cell.

You can always see the actual content of any cell in something called the ***formula bar***, which lives between the ribbon at the top and the spreadsheet (on a PC). Click on the cell, and the formula bar will tell you what's "really" in it.

Ribbon

B3 $f_x$ 14

| | A | B | C | D | E | F |
|---|---|---|---|---|---|---|
| 1 | | | | | | |
| 2 | | | | | | |
| 3 | | $14.00 | | | | |
| 4 | | | | | | |

Formula bar shows 14.

Cell B3 shows $14.00.

A few formats travel to the formula bar as well, such as percents and dates (which are actually numbers from Excel's point of view).

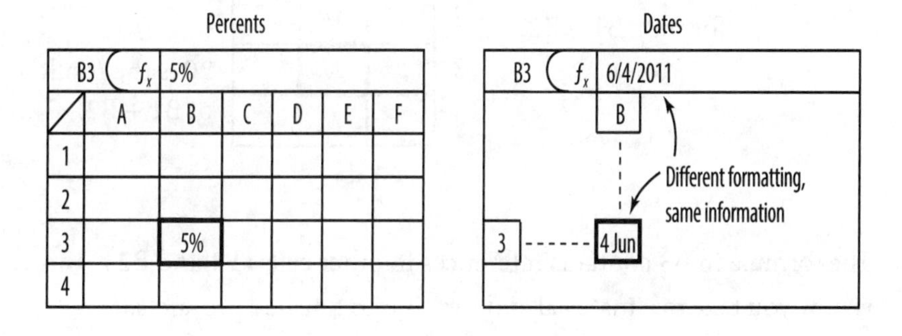


2. A cell can contain ***text***, obviously useful for labeling your model.

| | A | B | C |
|---|---|---|---|
| 1 | | | |
| 2 | | | |
| 3 | Rate | 5% | |
| 4 | | | |

A3 contains the text "Rate."
B3 contains the number 5%.

There's no technical connection between these two cells, but if you know what's good for you, you'll add labels as you build. Spreadsheets that only show numbers become cryptic to you and infuriating to your teammates.

3. A cell can contain a ***formula,*** which uses information from other cells to do a computation.

This is where it becomes really important to distinguish between what a cell *displays* and what it truly *contains.*

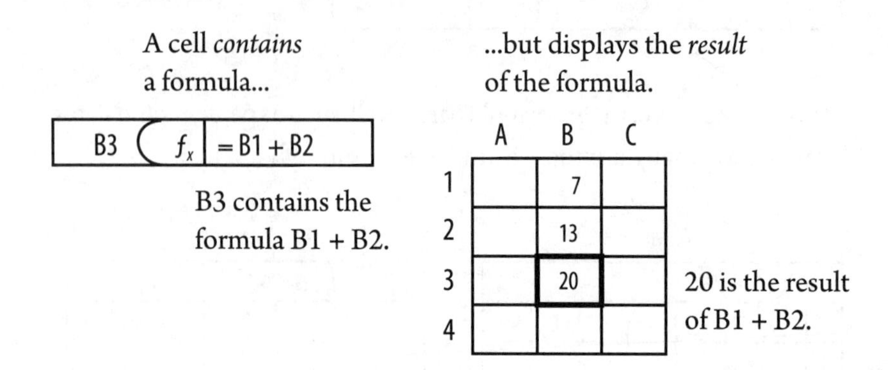


The formula in B3 contains references to other cells (B1 and B2). This is how you take raw material and turn it into finished products.

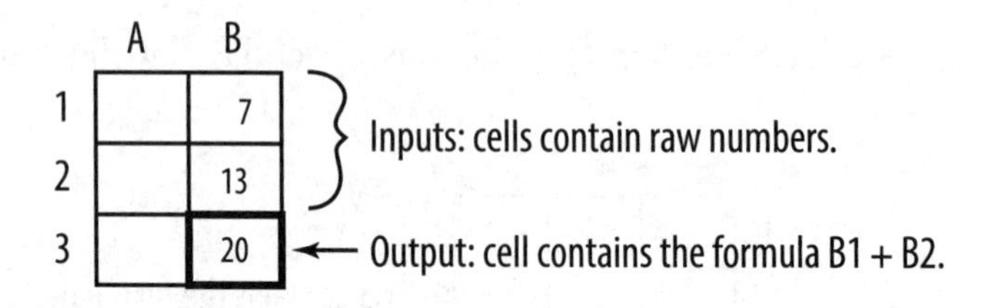


If 20 isn't your final output, then another cell can pick up that value and do something further with it. That cell would contain a formula with B3 in it.

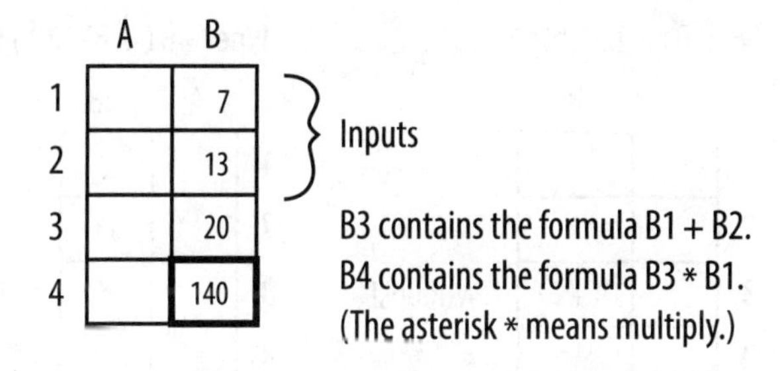


Normally cells do not display the formulas themselves, just the results of those formulas. So at a glance, you can't tell whether a number is ***hard-coded*** (you literally typed it in as a number) or ***computed*** by a formula. To reveal the true contents of a cell, you can highlight the cell and look at the formula bar.

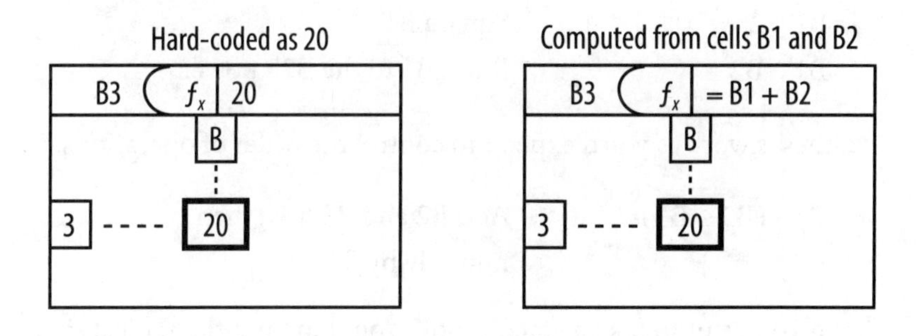


To tell Excel you want to put a formula in a cell, type an equals sign (=). You can also use a plus (+) or a minus (–) sign, but only under limited circumstances, so get used to starting with the equals sign. If you just type "B1 + B2" into a cell, Excel thinks you want that *text* in there.

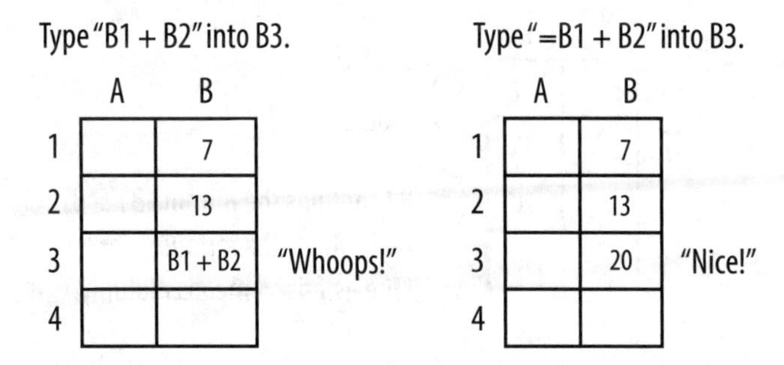


Excel can do all the common math operations:

| What you type into B3 | What you're telling B3 to do |
|---|---|
| =B1 + B2 | Add B1 and B2 |
| =B1 – B2 | Subtract B2 from B1 |
| =B1 * B2 | Multiply B1 by B2 |
| =B1 / B2 | Divide B1 by B2 |
| =B1 ^ 2 | Square B1 |
| =B1 ^ B2 | Raise B1 to the B2$^{\text{th}}$ power |

Parentheses work as you'd expect: to control the order of operations.

| | |
|---|---|
| =B1 * (B2 + B3) | Add B2 and B3 first, then multiply by B1. |

What if you want to take a square root? You don't use the radical sign (as in $\sqrt{9} = 3$). Instead, you use a pre-named ***function*** that Excel has defined for you. The function is SQRT.

| What you type into B3 | What you're telling B3 to do |
|---|---|
| =SQRT(B2) | Take the square root of B2 |

Functions always look like this:

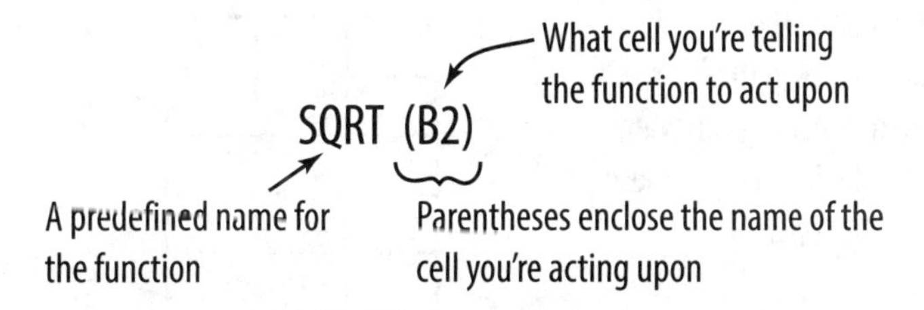


That format is similar to what you've seen in the math world: a function *f*(*x*) has the name "*f*," and it acts on *x* in some way (maybe it takes the square root). Excel just uses more descriptive names for its functions. You can find lists of all the functions (financial, statistical, etc.) by clicking on the *f*(*x*) button on the function bar: B3 $f_x$ ←

Some functions just take one cell as their ***argument*** (the input):

| What you type into B3 | What you're telling B3 to do |
|---|---|
| =SQRT(B2) | Take the square root of B2 |
| =FACT(B2) | Take the "factorial" of B2 |

Other functions take more than one cell as arguments. Just separate each argument with a comma.

| What you type into B3 | What you're telling B3 to do |
|---|---|
| =AVERAGE(B1, B2) | Take the average of B1 and B2, which is equal to (B1 + B2)/2 |
| =SUM(B1, B2) | Add B1 and B2 |
| =SUM(B1, B2, B3) | Add B1, B2, and B3 |

Where the functions AVERAGE and SUM really come into their own is with ***ranges***—whole rectangular groups of cells.

Name
B2:B5
"B2 to B5"—the colon is a "to" in this context, meaning you want Excel to be inclusive of these cells in the operation.

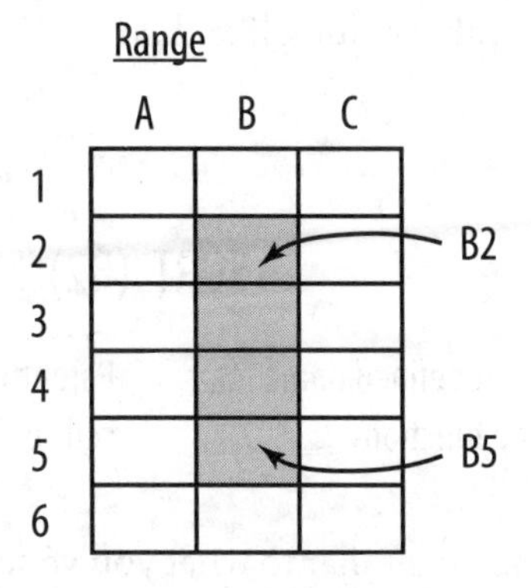


Name
B2:C4
Notice that B2 is the upper left corner, while C4 is the lower right corner.

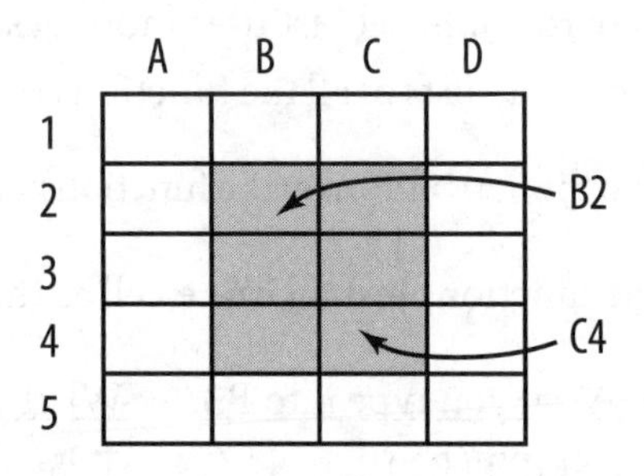


So now you can write powerful formulas like this:

| What you type into B3 | What you're telling B3 to do |
|---|---|
| =SUM(B2:B100) | Add up all the cells between B2 and B100, inclusive |
| =SUMPRODUCT(B2:B100, C2:C100) | Multiply B2 by C2, then B3 by C3, then B4 by C4, all the way down to B100 and C100, then add up all the results |

The SUMPRODUCT function is extraordinarily useful for computing the weighted-average return (or performance) of a portfolio of stocks, as in the example that follows.

| | A | B | C |
|---|---|---|---|
| 1 | Stock | Return | Weight |
| 2 | IBM | 5% | 1.2% |
| 3 | AT&T | −1% | 0.8% |
| 4 | Google | +10% | 0.5% |
| 5 | Microsoft | −3% | 1.5% |
| | ⋮ | ⋮ | ⋮ |
| 99 | Exxon Mobil | +3% | 0.9% |
| 100 | Du Pont | −2% | 0.6% |

SUMPRODUCT(B2:B100, C2:C100)
= (5% × 1.2%)
+ (−1% × 0.8%)
+ (+10% × 0.5%)
+...
+ (−2% × 0.6%)
= total return of portfolio

As you work with formulas and functions, you'll notice that the references to input cells and ranges show up outlined in color—and the cells and ranges themselves are temporarily displayed with those colors as well. This helps you make sure you're adding or "sumproduct-ing" the right cells. In fact, to edit the formula, you can grab the colored highlighting on the worksheet itself and move it around.

What if you wanted to add up a range of cells and take the square root? You can do this in two steps.

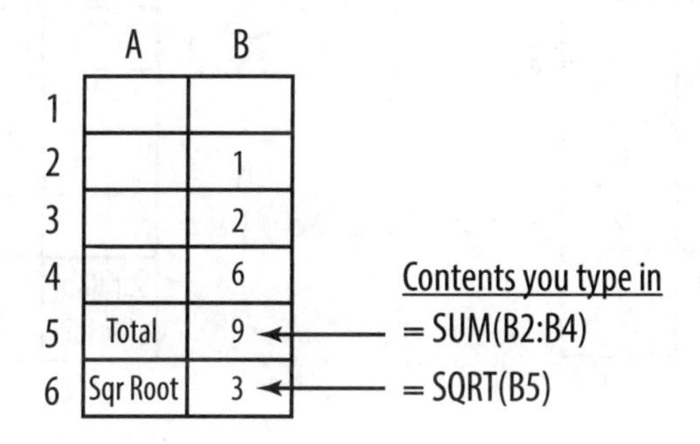

Or you could do it in one step with a ***nested*** or embedded formula.

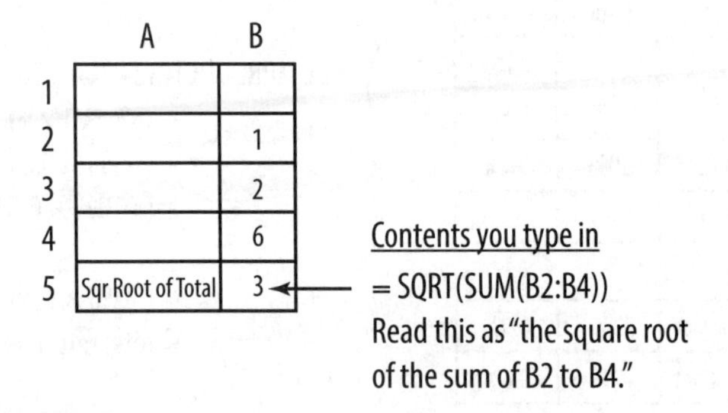


Which should you prefer? Some Excel jockeys love nested formulas. Our advice? Avoid them (the formulas, not the jockeys) as much as you can. Nesting quickly becomes inscrutable to you and to everyone else who has to follow the logic of your model. Let Excel do what it's great at—showing the innards. You can always hide rows or columns for a presentation.

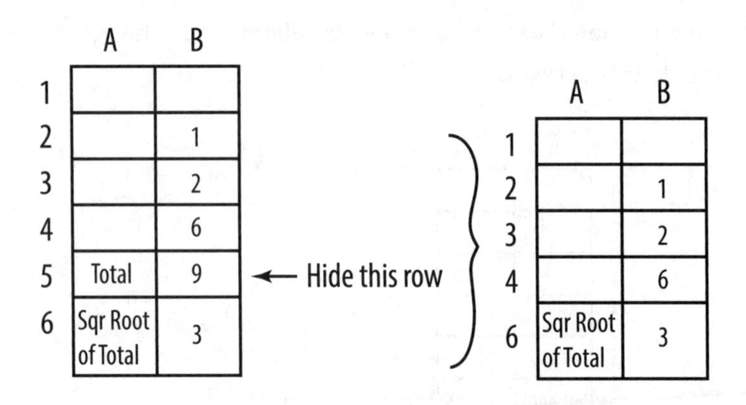


Likewise, avoid hardcoding input numbers *inside* formulas and functions. Keep your work transparent and adjustable by putting inputs in separate, visible cells—shaded in sunburst yellow.

**Bad**

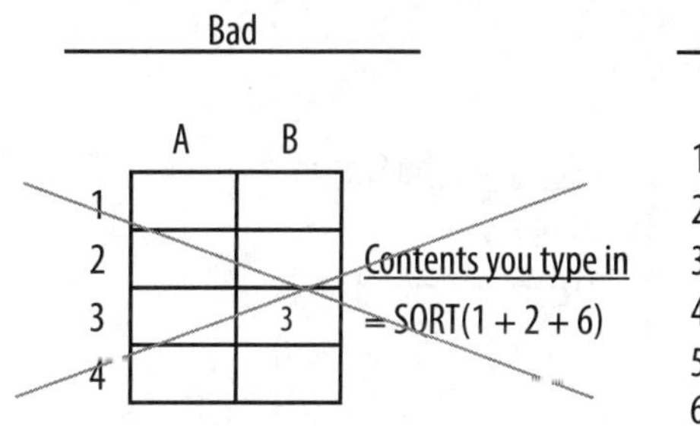


- Completely hidden work
- Inflexible—you have to go inside the function to change anything

**Good**

| | A | B | Contents you type in |
|---|---|---|---|
| 1 | | | |
| 2 | | 1 | |
| 3 | | 2 | |
| 4 | | 6 | |
| 5 | Total | 9 | = SUM(B2:B4) |
| 6 | Sqr Root | 3 | = SQRT(B5) |

- Open layout
- Flexible—you can change the output by editing the contents of B2, B3, or B4

## IT'S NOT ALWAYS RELATIVE

Say you have a big column of data, and you want the square of every number in that column.

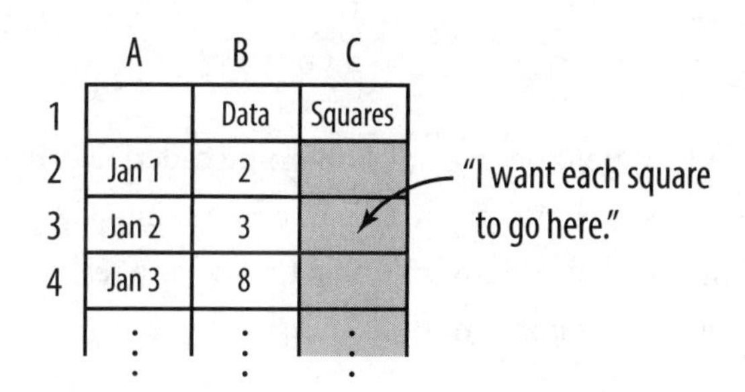


All you have to do is write *one* good formula and copy it.

| | A | B | C |
|---|---|---|---|
| 1 | | Data | Squares |
| 2 | Jan 1 | 2 | 4 |
| 3 | Jan 2 | 3 | |
| 4 | Jan 3 | 8 | |
| ⋮ | | | |

Contents you type in

= B2^2 (for cell C2)

(that's B2 raised to the 2nd power, or B2 squared)

What happens when you copy the content of C2 down the column? Amazingly, the formula adjusts to find the right cell—the cell immediately to the left.

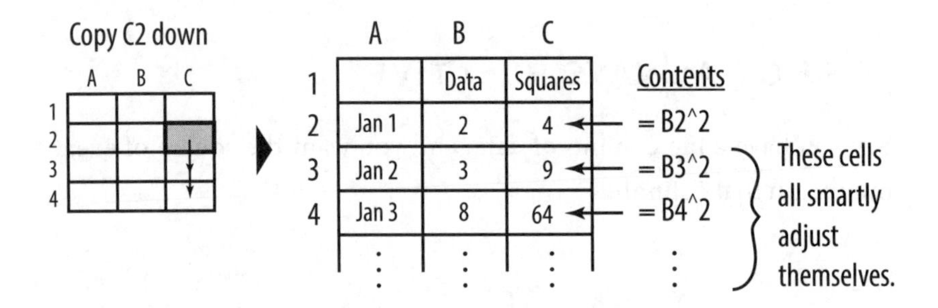


That adjustment happens because of something called ***relative addressing*** or ***relative references.*** When you typed "=B2^2" into C2, you *thought* you were telling Excel to square B2. In fact, you were really telling it to square whichever cell is ***one space to the left***.

| | A | B | C |
|---|---|---|---|
| 1 | | Data | Squares |
| 2 | Jan 1 | 2 | 4 |
| 3 | Jan 2 | 3 | |
| 4 | Jan 3 | 8 | |

In cell C2, the formula "=B2^2" *really* means "square the cell to my left."

Excel assumes you want relative addressing, because it makes big spreadsheets much easier to build. The opposite of relative addressing is ***absolute addressing***, which means "refer back to such-and-such a cell, no matter where I copy this formula." Absolute addressing is useful when you're computing percents of a whole. You use dollar signs ($) in front of the column letter and row number to lock the cell reference absolutely.

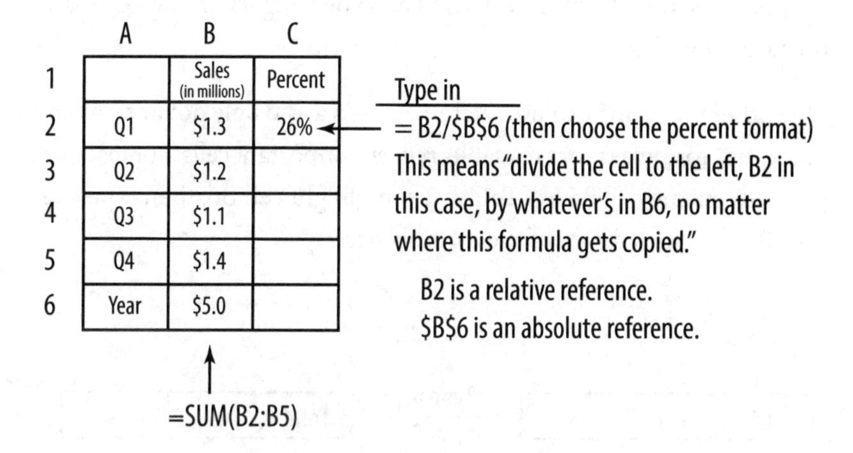


Watch how the formula changes as you copy it down. The ugly dollar signs keep the second part locked.

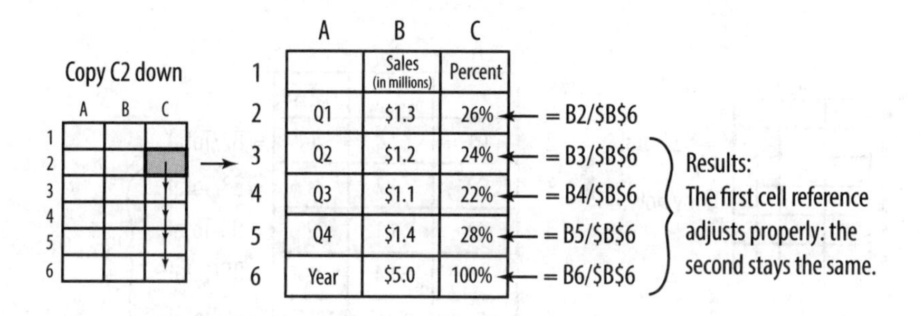


If you have "=B2/B6" in C2 by mistake and then copy down, you get "=B3/B7" and other incorrect formulas, so be careful.

Play with relative and absolute addresses on a scratch worksheet by copying simple formulas around. You'll see how dollar signs keep cell reference locked, while relative addresses (no dollar signs) adjust to each new position. If you're in the formula bar, you can use the F4 function key to toggle between relative and absolute addressing, instead of typing in dollar signs (which also works). By the way, function keys in Excel are pretty useful. The more you can keep your fingers on the keyboard, the faster you'll go.

One other point on absolute references—to avoid ugly dollar signs and make your formulas more readable, ***rename*** important cells. For instance, you can rename cell B6 as "Total." Just highlight cell B6, then type over the "B6" on the left side of the formula bar:

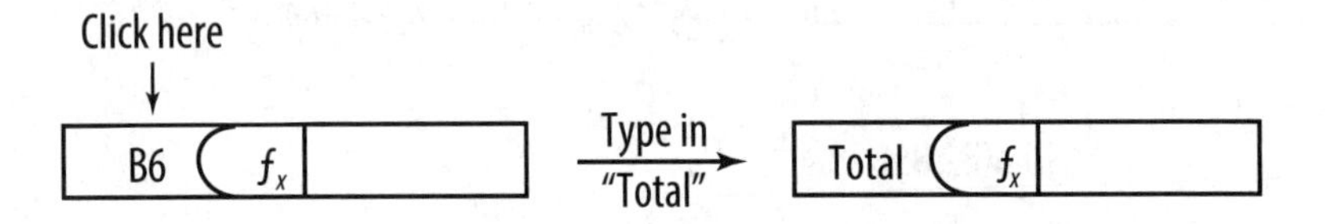


Now you can refer to B6 in formulas as "Total." That's an absolute reference that will never change, so only rename important cells.

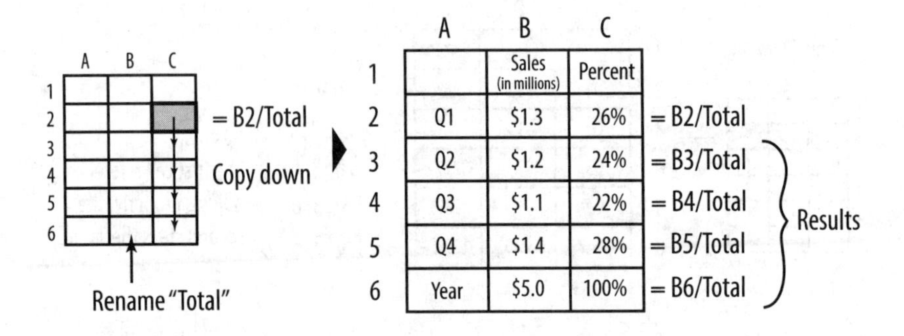

## GOAL SEEK AND SOLVER

Remember the four components of a model?

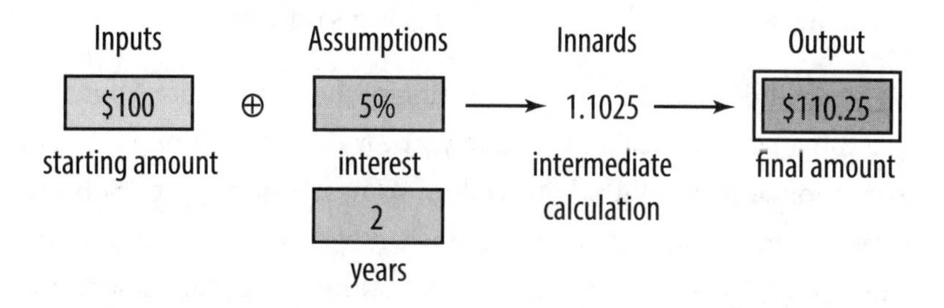


To get $200 in two years, how much money do you need right now? You have four ways to figure this out:

1. Build a brand new model. Ugh.
2. Do the math outside of Excel. Double ugh.
3. Plug various inputs in on the left until you get $200 out on the right. Feasible but a pain.
4. Tell Excel to plug different inputs in on the left until you get $200 out on the right. Much better.

This lifesaving Excel tool is ***Goal Seek***—you tell Excel to seek a goal.

Click on your output cell to highlight it, then choose Data → What-If Analysis → Goal Seek. You'll get a little dialog box:

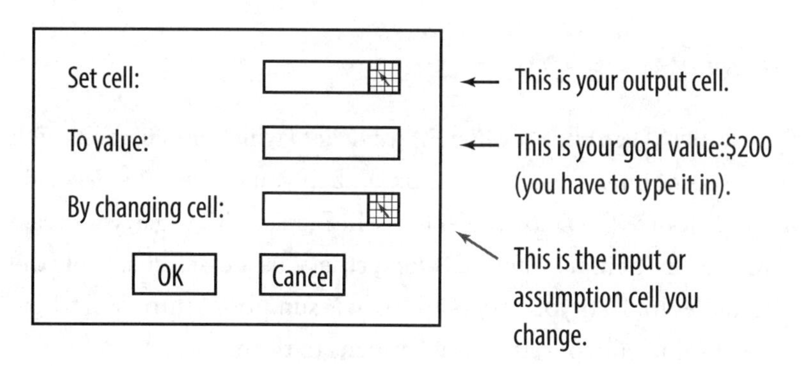

Type in cell references, or click to insert them, then hit OK. Excel plugs a bunch of numbers in, faster than the eye can see, and proposes an answer. In this case, the input that works is $181.41 (rounding). The bigger your model, the more pain the Goal Seek tool can save you.

***Solver*** (an add-in on the Analysis section of the Data ribbon) does the same thing as Goal Seek, but it lets you vary more than one input or assumption at once. With that freedom comes responsibility. Solver is a very powerful tool, but you have to use it wisely. For instance, in the model above, if you try to vary both assumptions at the same time to find a particular goal, Solver goes off the rails because many combinations of years and interest rates give you the result you want, and Solver doesn't know which combination to pick.

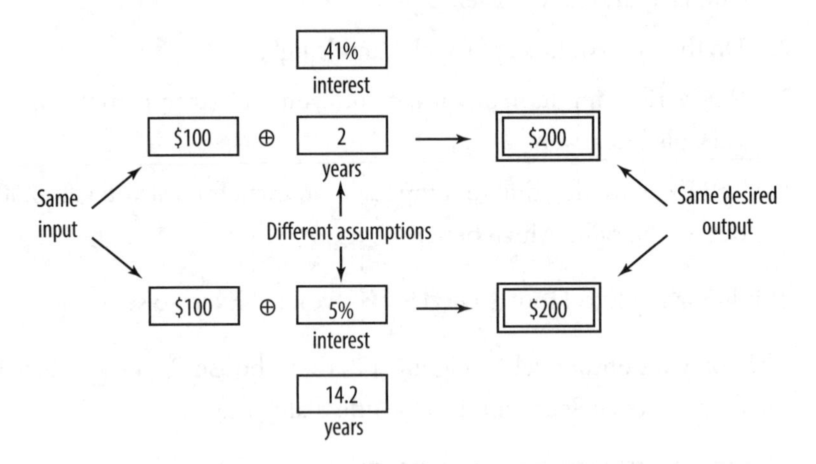


## WHAT'S THE SCENARIO?

If you want to keep track of different ***scenarios*** (which are situations with whole sets of inputs and assumptions), then you can use the ***Scenario Manager*** tool or the ***Data Tables*** tools. In Scenario Manager, you create combinations of inputs labeled "worst case," "best case," etc., while the Data Tables tool lets you vary one or two assumptions through a whole range of values and observe what happens to the output.

<u>Scenario Manager</u>

| Inputs | Scenario |
|---|---|
| 5% interest<br>2 years | Worst case |
| 10% interest<br>4 years | Best case |
| 7% interest<br>3 years | Likely case |

<u>Data Tables</u>

| Interest | Output |
|---|---|
| 5% | $110.25 |
| 6% | $112.36 |
| 7% | $114.49 |
| 8% | $116.64 |
| ⋮ | ⋮ |

Finally, an add-in program called ***Crystal Ball (CB)*** lets you run *thousands* of scenarios in Excel and aggregate the results. Crystal Ball generates the scenarios at random, but according to guidelines you give it. For instance, you might tell Crystal Ball to pick interest rates between 5% and 10%, with any interest rate equally likely...

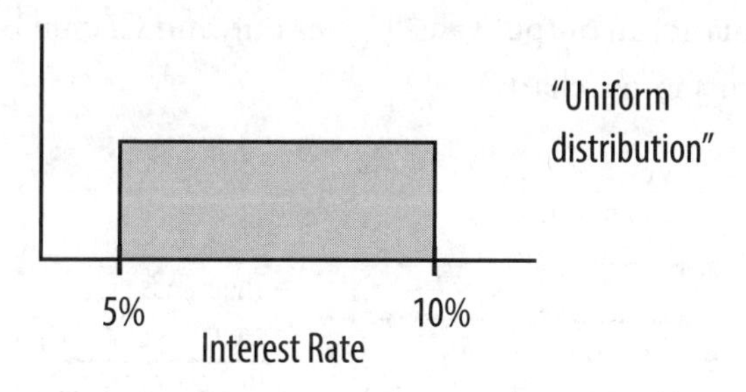


...or with a peak of some kind around 7.5% so that more scenarios with interest rates close to 7.5% occur.

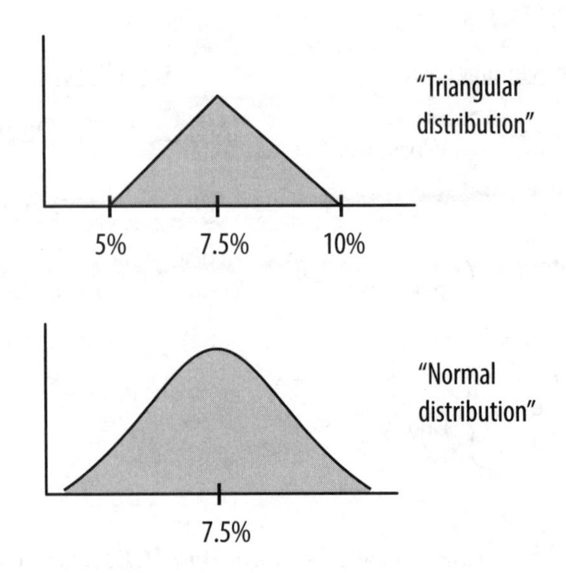


For more on distributions, see Chapter 3: Statistics.

Crystal Ball generates its many thousands of scenarios by rolling the dice, so to speak, over and over again according to your casino rules. For each scenario, an output value comes out, and Crystal Ball collects all those results into a chart.

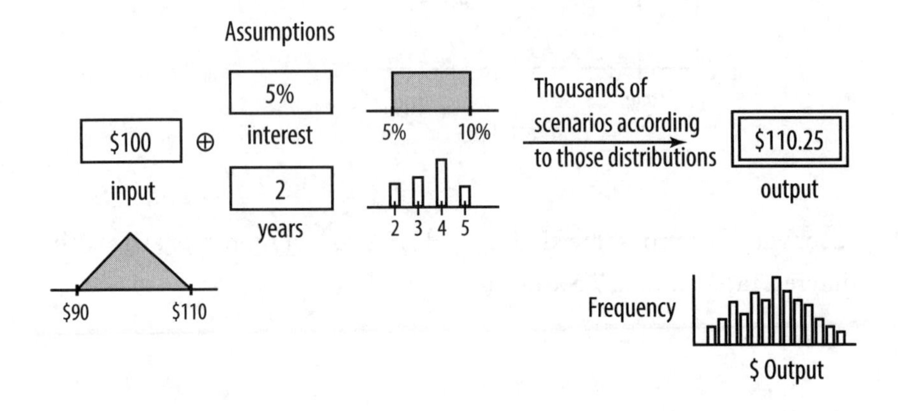

That frequency chart of outputs (or histogram) can tell you a lot—what your average output is, how likely you are to make over $120, etc.

Yet again, Spidey, with great power comes great responsibility. Crystal Ball is only as good as its assumptions. Without any further direction from you, CB assumes that all the random numbers in the scenarios are independent of one another, like coin flips. That may or may not be true. For instance, housing prices throughout the US didn't all move in the same direction very often until, one day, they did. Despite the name, the power of Crystal Ball (and, more broadly, Excel) to predict the future is limited, and it's dangerous to forget those limits. In fact, what Crystal Ball does is called ***Monte Carlo analysis***, after the famous European gambling destination. Never forget you're spinning a roulette wheel with CB.

## DATA MANAGEMENT

You can also keep track of lots of data in Excel. Just be sure to organize the data in a clean table:

| | A | B | C | D |
|---|---|---|---|---|
| 1 | First Name | Last Name | Date | Purchase |
| 2 | Anne | Bancroft | 7/10/18 | $100 |
| 3 | Clark | Gable | 7/12/18 | $150 |
| 4 | Anne | Bancroft | 7/13/18 | $70 |

Each row should represent a separate ***record***—a purchase, for instance. Each column should represent a separate ***field***—some bit of data you want to track about each purchase (customer first name, customer last name, etc). Make sure that there are no gaps in the table and that it's not touching anything else. Put field names in the first row.

Excel is not truly a "database" program, but if you set up the table this way, you can do a lot:

- ***Sort*** by any field (say, from earliest to latest)
- ***Filter*** by field (say, you just want to look at purchases on a particular date)
- Create ***subtotals*** by field
- Create ***pivot tables***

Pivot tables are super-flexible tools for exploratory data analysis. Imagine that you're a store owner, and you have a big list of purchases all scrambled up. You have a lot of repeat customers, even on the same day. You can total up the purchases by customer *and* by day at the same time, using a pivot table. It's a jiffy to set up. Just click the Pivot Table button on the Insert ribbon, hit OK, then drag the field names into position.

Purchases → Pivot Table of Purchases

| Full Name | Date | Purchase |
|---|---|---|
| Anne Bancroft | 7/10/18 | $100 |
| Clark Gable | 7/12/18 | $150 |
| Anne Bancroft | 7/13/18 | $70 |
| Anne Bancroft | 7/11/18 | $80 |
| Sam Spade | 7/12/18 | $100 |
| Clark Gable | 7/11/18 | $50 |
| Clark Gable | 7/12/18 | $25 |
| Anne Bancroft | 7/12/18 | $75 |
| Sam Spade | 7/11/18 | $100 |
| Sam Spade | 7/10/18 | $75 |
| ⋮ | ⋮ | ⋮ |

| Full Name | 7/10 | 7/11 | 7/12 | | Subtotal |
|---|---|---|---|---|---|
| Anne Bancroft | $100 | $150 | $75 | ··· | $1,000 |
| Clark Gable | $0 | $50 | $175 | ··· | $1,250 |
| Sam Spade | $75 | $100 | $100 | ··· | $750 |
| | ⋮ | ⋮ | ⋮ | | ⋮ |
| Subtotal | $1,200 | $400 | $500 | | $10,000 |
| | | | | | Total |

## LAST THOUGHT

If Excel starts to beat you down, do one simple thing—remove gridlines from the display. Go to the View ribbon and uncheck gridlines.

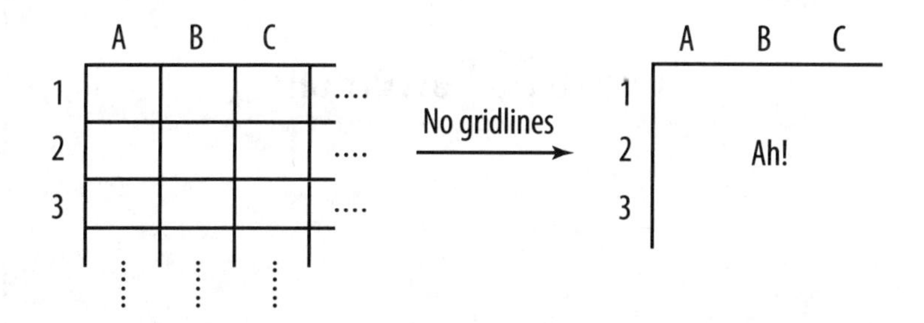


Suddenly Excel looks like a blank sheet of paper. If you're more a Word person than an Excel person, the visual shift will be surprisingly uplifting.[3]

3 One of us sat down with her study group for the first time to tackle a Corporate Finance project. The other three team members (all of whom were guys named Mike) opened up their PCs and launched Excel. The author cheerfully flipped open her MacBook and began typing into a Word table. Finally someone mentioned the F4 key. Needless to say, this author could have used an Excel primer before school.

How to Do PowerPoint

Agenda

1. What Is PowerPoint?
2. Principles of Good PowerPoint
3. Principles in Action

## What Is PowerPoint?

- Most common presentation software
  - Projected on-screen
  - Printed for handouts
- Based on slideshow analogy

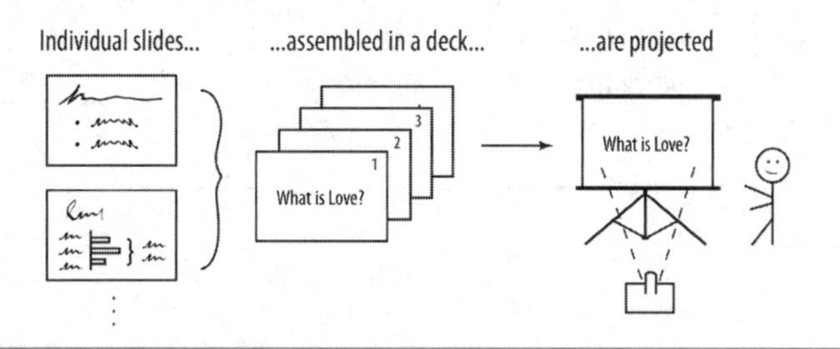


## Agenda

1. What Is PowerPoint?
2. Principles of Good PowerPoint
3. Principles in Action

# Principles of Good PowerPoint

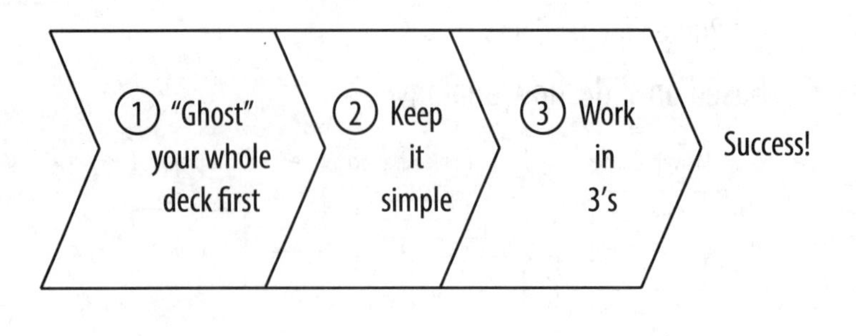


# 1. "Ghost" Your Whole Deck First

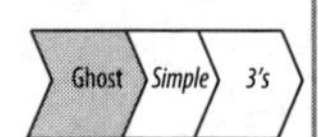


a) Draw 9 squares on a blank sheet of paper.

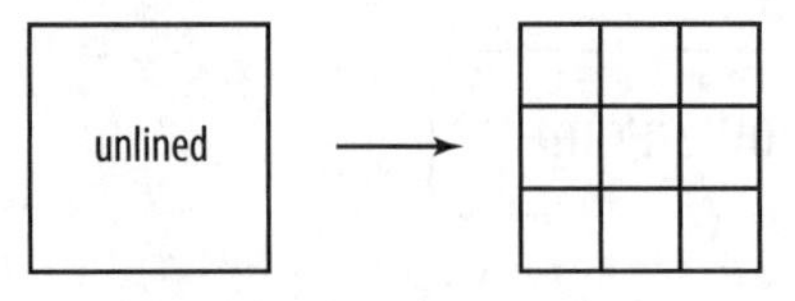


b) Put 1 big message per slide and "ghosts" (placeholders) for details.

| Title | Agenda | Mkting is Strong $X $X $X |
|---|---|---|
| Operations are Weak | Morale is High pie chart | |
| | | |

Focuses you on the overall storyline.

## 2. Keep It Simple

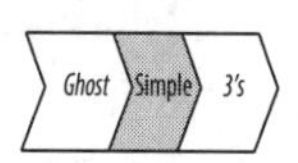


**Whole Deck**

- Use clear signposts
  - Agenda slides

  - Trackers in corners

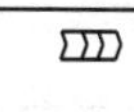

**Each Slide**

- Fewer words = better
- Large enough for the back row
- Modest animations/transitions

## 3. Work in 3's

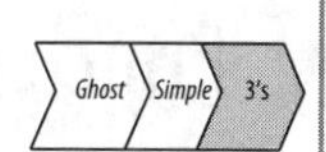


Government

- of the people
- by the people
- for the people

- Memorable balance
- 2's and 4's work as well
  - Don't be *too* predictable

# Agenda

1. What Is PowerPoint?

2. Principles of Good PowerPoint

3. Principles in Action ⟵ This very presentation!

# Principles in Action—Ghosting

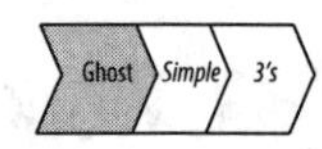


Actual Ghost of This Deck

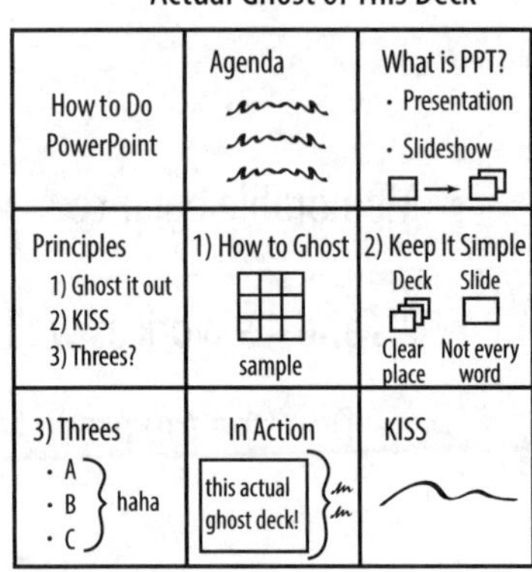


- Done with pencil and paper
- Took only a half-hour subway ride
- Prevented "story creep"

## Principles in Action—Simple

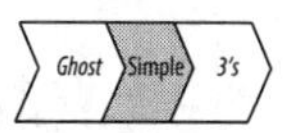


- Clear signposts throughout

Slide

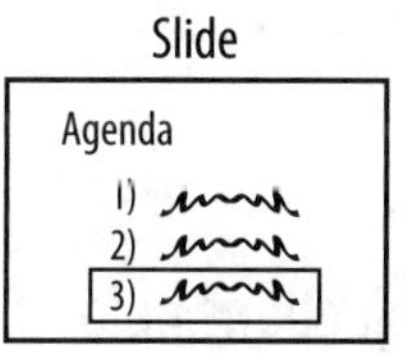


Slide

Tracker for principles

- Few words

Slide

- Clear signposts throughout
- Few words

## Principles in Action—3's

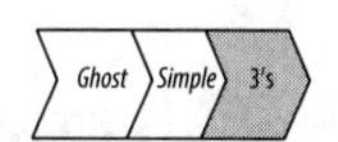


- Three items on agenda
- Three principles of good PowerPoint
- Three bullets, in most places
  - Sometimes two
  - Not averse to four

## Can You Ever Deviate from Plan?

- Of course you can.
- Try to anticipate objections beforehand.
  - Plan your responses and work them into the deck.
  - Or have "hip-pocket" extra slides ready.
- During the presentation, deal with questions as they arise.
  - Audiences hate waiting till the end.

## Final Thoughts

- Recap the big takeaways—the short "so-what's."
  - Any that you planned to make
  - Any that came out of the discussion
- Lay out next steps.

# And Don't Do This...

Swooshing animations!!!

Crazy transitions!!!

Crowded tiny text that no one can read from the back row so that you have to say, "Sorry, I know you can't read this, but..." That's terrible. This includes graphs with tiny labels.

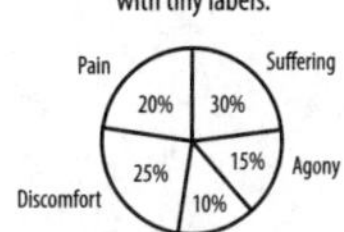


The entire text word-for-word of what you're going to say, so you stare at the screen and read it aloud to your audience as if they were five years old.

Clip art unless you are a legitimate graphic artist (and you're probably not)

# ...Do This Instead

Nothing

# Chapter 2: Economics & Game Theory

This chapter has three major sections:

1. ***Microeconomics***: "small-scale" economics of individual buyers and sellers
2. ***Macroeconomics***: "large-scale" economics of nations and governments
3. ***Game theory***: the study of certain formalized "games" between players

Depending on your program, these topics may be dealt with in more than one course. They may also use different names, such as "Managerial Economics" or "Global Markets," just to sound cool.

## MICROECONOMICS

Microeconomics attempts to describe the ***physics of markets***,[1] where buyers come to buy what sellers have to sell. Somehow a ***price*** gets set and a ***quantity*** of the product gets sold. The buyers and sellers are assumed to be acting ***rationally*** (i.e., intelligently) and in their own ***self-interest***—that is, each one wants to get the best deal for himself or herself. That might not always be 100% true, but the ***"neoclassical"*** version of microeconomics, which is what you'll be taught, assumes that everyone is smart and selfish in a marketplace. It's often a good assumption:

1 Yes, one of us used to teach high-school physics. Is that obvious?

just think of the New York Stock Exchange, where many intelligent, self-interested investors bust tail every day to eke out thin advantages. Competitive financial markets such as those for US "large-cap" stocks (those of big companies) are considered relatively ***efficient***—most advantages are temporary, as information rockets around the investing community.

## DEMAND CURVE

A lot of microeconomics is done by drawing lines and curves on graphs. A single curve (we'll use this term whether the line is actually curved or straight) can represent a variety of real-world situations.

For instance, a ***demand curve*** captures the following two scenarios:

1. At a high price, only a few people want to buy.
2. At a low price, lots of people want to buy.

This is shown by plotting the price $P$ on a vertical axis and the quantity $Q$ that people demand on a horizontal axis. ($Q$ might be written as $Q_D$, where the subscript $D$ means "demand.")

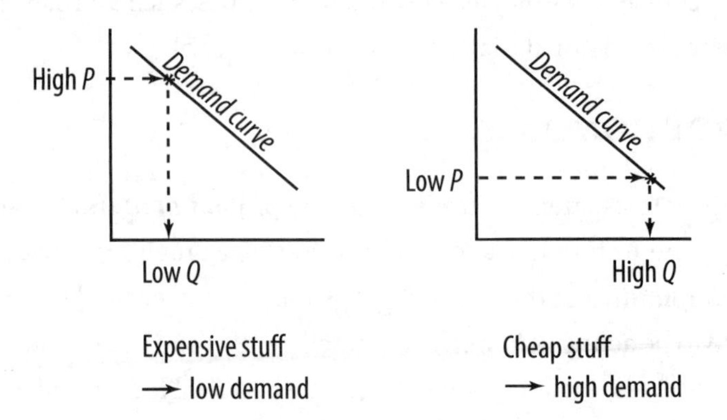


For some historical reason, even though $P$ is the "independent" variable, it goes on the vertical axis. Quantity $Q$, which depends on $P$, goes on the horizontal axis. Usually graphs go the other way. When you see a curve plotted on $x$ and $y$ axes, the $x$ variable is what typically determines $y$.

But here, it's *P* that determines *Q*, most logically. Not a big deal, but it's worth pointing out.

Imagine that at a price of $10, the buyers demand 200 units of whatever the product is. If those 200 units are sold at that price, then the sellers have collectively made ($10 per unit) × (200 units) = $2,000. That $2,000 is ***revenue R* = *PQ***. On a demand curve, revenue is the area of a rectangle.

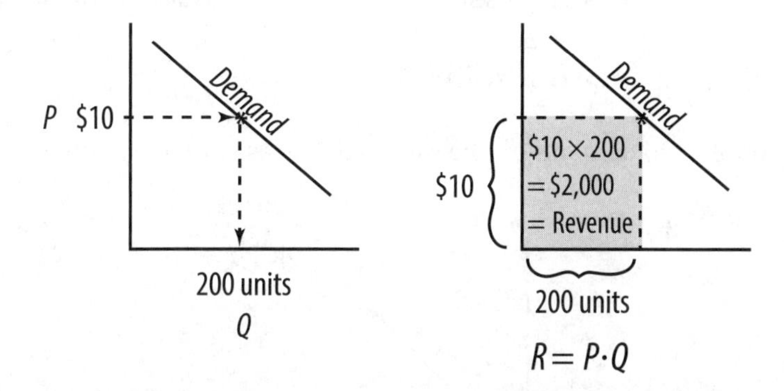


By the way, sometimes prices are written just in dollars ($10), but you should always be ready to think of them as dollars ***per unit*** ($10/unit). This will help you remember that you multiply per-unit prices by units (quantity) to get a total amount in dollars (revenue in this case).

## SUPPLY CURVE

A demand curve captures how much buyers demand at various prices. In contrast, a ***supply curve*** captures the behavior of sellers: how much they're willing to sell at various prices, or how many sellers there even are. Simply put, lots of people want to sell at a high price. On the other hand, if prices are low, only a few people want to sell. (On this graph, *Q* can be written as $Q_S$ for "quantity supplied.")

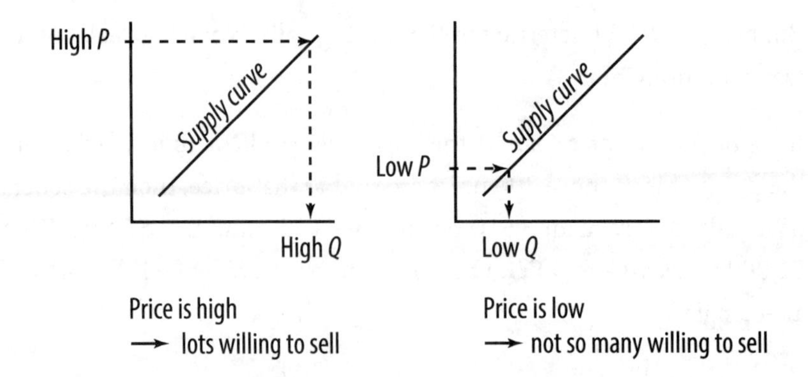


## SUPPLY AND DEMAND

Put supply and demand on the same set o' axes, and presto, they cross.

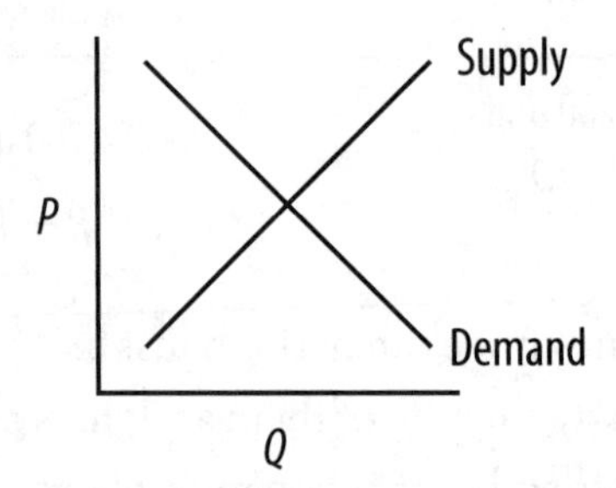


The magical intersection point represents a state of ***equilibrium*** or balance. At that point, supply equals demand. Every buyer who wants to buy does so; every seller who wants to sell does so. The *market clears,* as they say—there are no unfulfilled desires at the equilibrium price, or ***market price***.

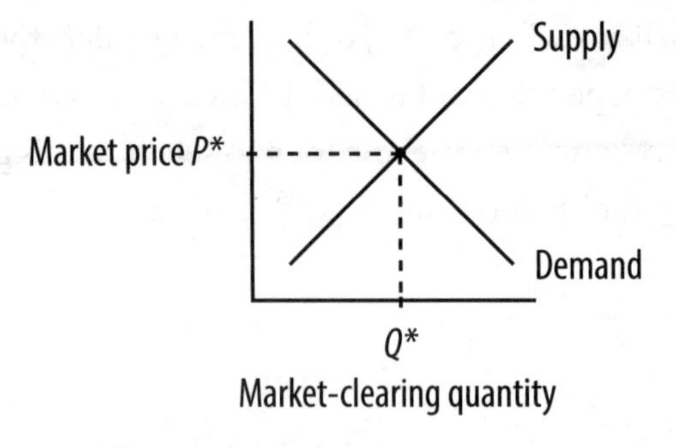

You always assume that you have ***perfect competition***, which means having lots and lots of buyers and sellers, none of whom can corner the market. Under these conditions, supply and demand converge at a specific price and quantity.

If the price is momentarily too high, more quantity is supplied than demanded. Buyers are scarce, so sellers cut their prices to stimulate demand.

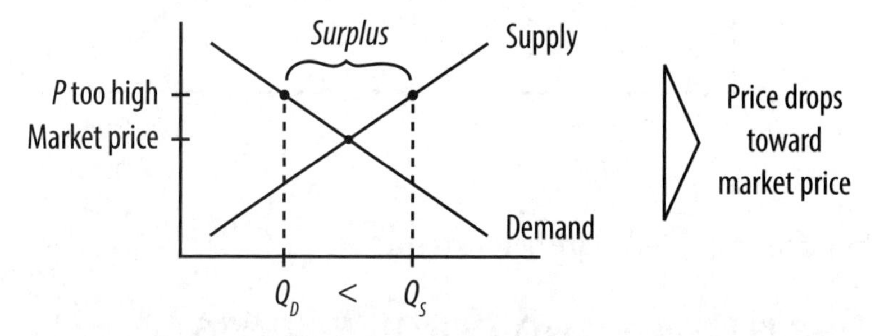


On the other hand, if the price is momentarily too low, more quantity is demanded than can be supplied. Sellers are scarce, so buyers offer higher prices to encourage sales:

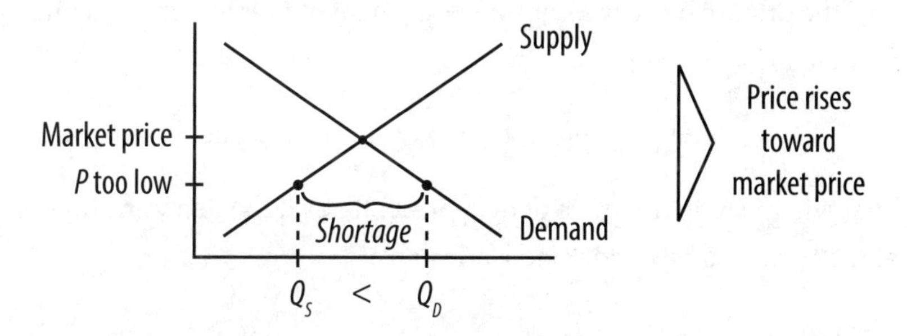


Thus, the ***market mechanism*** creates a stable equilibrium, in theory.

You can model various "shocks" to the system with these curves. For instance, the entrance of a new supply of product pushes the supply curve to the right, leading to a lower market price at a greater quantity sold. This should make sense. In general, when you move these curves around, the results should not be surprising. Imagine a real-world scenario (or even a fictional one). Anyone who's watched HBO's classic show *The Wire* can picture what happens to the market price and quantity when a new supply of heroin hits the street:

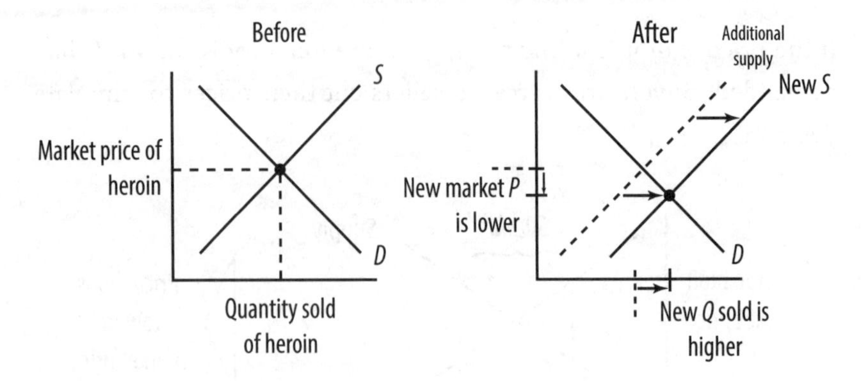


The price drops, and more heroin is sold.

## SUBSTITUTES AND COMPLEMENTS

***Substitute*** products, like butter and margarine, can replace each other. Do you care whether you're using one or the other? Probably not.

So if the price of butter goes up, lowering demand for butter, the demand for margarine goes up:

$$\text{If } Q_D \text{ for butter} \downarrow \text{ then } Q_D \text{ for margarine} \uparrow$$

That will push up the price of margarine. It's as if the demands for each of the two products add up to a constant:

$$\underset{\text{(butter)}}{Q_D} + \underset{\text{(margarine)}}{Q_D} = \text{some overall demand for yellow fatty spreads}$$

On the other hand, products that are ***complements*** go together and aren't worth as much without the other. Think about butter and toast...

With complementary products, the demands move in the same direction. So if the price of butter goes up, lowering demand for butter, the demand for toast also goes down:

$$\text{If } Q_D \text{ for butter} \downarrow \text{ then } Q_D \text{ for toast} \downarrow$$

This will push down the price of toast. The difference between the demands is always the same:

$$\underset{\text{(butter)}}{Q_D} - \underset{\text{(toast)}}{Q_D} = \text{a constant}$$

## ELASTICITY

Elasticity answers the following questions:

- If you change the price a bit, how much does demand change?
- Under the same conditions, how much does supply change?

Elasticity is a measure of how much supply and demand stretch in response to price shifts. Think of a rubber band that you hang a weight from. The more elastic the rubber is, the farther the weight hangs down.

When you are trying to determine how elastic supply or demand is, you usually increase the price by 1% and watch what happens to the quantity. This way, you can compare the elasticities of $20 haircuts and $20,000 cars. Likewise, the change in quantity is measured as a percentage.

***High elasticity*** = big response to price shifts = shallow slopes

A 1% increase in price shifts quantity by 5%.

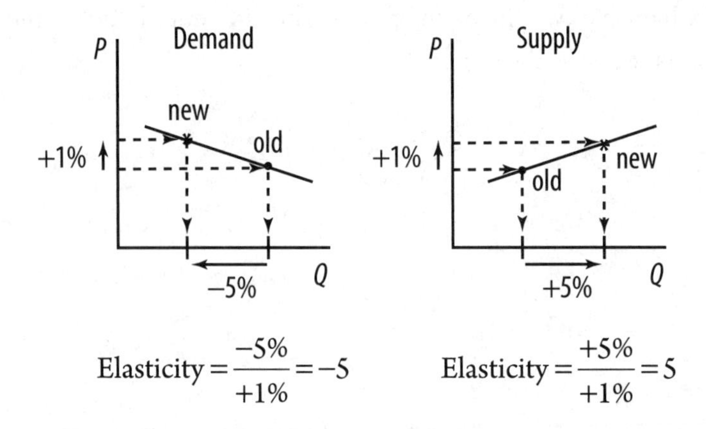


$$\text{Elasticity} = \frac{-5\%}{+1\%} = -5 \qquad \text{Elasticity} = \frac{+5\%}{+1\%} = 5$$

***Low elasticity*** = small response to price shifts = steep slopes

A 1% increase in price shifts quantity by only 0.2%.

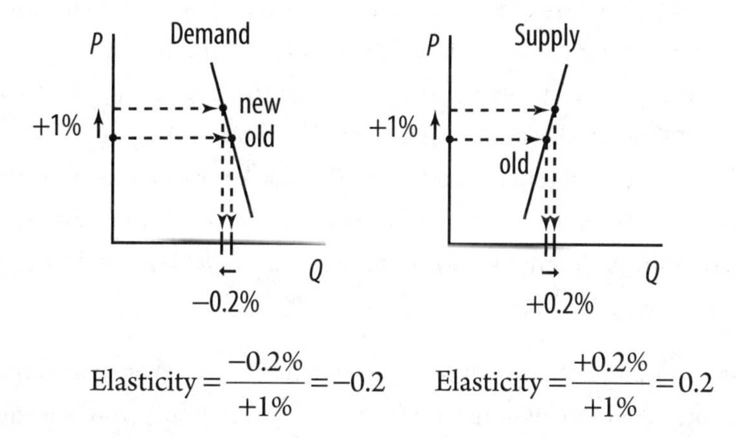


The formula for elasticity is this:

$$E = \frac{\%\Delta Q}{\%\Delta P} \quad \text{("percent change in } Q \text{ over percent change in } P\text{")}$$

Percent change in $Q$ is the actual change in $Q$ divided by $Q$ (expressed as a percent). The $\Delta$ ("delta") means "change in" something, so $\%\Delta Q = \frac{\Delta Q}{Q}$ and $\%\Delta P = \frac{\Delta P}{P}$.

If the result is greater than 1 or less than −1, then demand or supply is considered highly elastic. If the result is between −1 and 1, then demand or supply is considered inelastic.

## DEVIATIONS FROM PERFECT COMPETITION

One way the market mechanism can be distorted is through ***cartels*** of sellers or of buyers. A cartel of sellers bands together to limit supply and keep the price high. Likewise, a cartel of buyers tries to limit demand to keep the price low.

If supply is truly limited *and* a cartel of sellers can enforce its policy, then the scheme might work. For instance, the oil producers' cartel, OPEC, seems to do just fine. But cartels are ***anti-competitive***, so they can run afoot of ***antitrust*** laws in the United States. Moreover, cartels are unstable, because every member of the cartel has a strong incentive to cheat. For example, if you're an oil-producing country, you can make piles of money if you convince everyone *else* to obey the cartel. Overall supply is limited, prices stay high, and then you secretly sell more oil on the side. Cartels fly in the face of the market mechanism, so they always have to be enforced by fundamentally non-economic means (e.g., threats of violence). Again, shows such as *The Wire* and *Breaking Bad* are instructive here.

Taking anti-competitive behavior even further, you can have a ***monopoly*** (just one seller) or ***monopsony*** (a much less common word meaning "just one buyer," and not a mutant Milton-Bradley game). If you have a monopoly, you're your own cartel. It's generally going to be in your interest to limit supply and keep prices high, all else being equal. Monopolies are typically illegal as well, unless there's a ***natural monopoly***, in which the characteristics of production (such as enormous fixed costs or long-term contracts between any buyer and the seller) push the market toward having just one seller. Think electric utilities—at the end of the day, it only makes sense to run *one* power line to your house, and building a power plant costs loads of dough. When a market has the characteristics of a natural monopoly, government regulators may step in, as they do with utilities, to set prices and quantities.

In contrast, the US government has a natural monopsony for certain kinds of military equipment produced by domestic manufacturers, such as Lockheed Martin. The US is the only (legal) buyer of these products, thereby dictating the overall demand.

***Taxes*** and ***quotas*** (artificial limits on quantity, either minimums or maximums) also distort markets, meaning that the resulting $P$ and $Q$ are not the theoretical $P$ and $Q$ you'd have without the taxes or quotas. That's not a value judgment; the taxes or quotas may serve a greater good, or they may not. Either way, they impact the market.

Finally, some markets have ***network externalities***, meaning that buyers don't act independently. With ***positive*** network effects, buyers want the product *more* when there are more buyers; the product is worth more to the millionth buyer than to the first. Facebook, the fax machine, and the telephone all work this way, since the more people you can friend, fax, or call, the better for you. In contrast, products with ***negative*** network externalities are worth *less* when more people use them. Think of anything exclusive, anything with snob appeal—as soon as the masses get the new Louis Vuitton bag, it's no longer cool. Likewise, overcrowding can diminish your enjoyment of an amusement park (or your broadband connection).

## COST CURVES

And now for something completely different... cost curves.

Before diving in, you should know that *cost* in microeconomics is not quite the same things as *cost* in accounting, though the concepts overlap. Cost in micro is forward-looking—specifically, you need to remember to consider ***opportunity costs*** (the benefits of paths you're not choosing) and ignore ***sunk costs*** (the pain of past choices), when you're making a rational economic decision.

Here's a rogue's gallery of cost variables you might plot on a graph (usually against quantity Q's on the horizontal axis).

| | | |
|---|---|---|
| ***Total Cost*** | ***TC*** | The total cost of making Q units |
| ***Fixed Cost*** | ***FC*** | The fixed constant cost you incur, whether or not you make any units at all—say, the cost of building a plant |
| ***Variable Cost*** | ***VC*** | The cost that varies with the number of units you produce—e.g., raw materials costs |

These first three variables are related by a simple sum:

$$TC = FC + VC$$

For instance, a simple formula for $TC$ as a function of quantity $Q$ might be this:

$$TC = 800 + 2Q$$

Plugging in various values for $Q$, you can see the total cost of making various quantities:

| $Q$ | $TC = 800 + 2Q$ |
|---|---|
| 0 units | $800 + 2 \cdot 0 = \$800$ |
| 100 units | $800 + 2 \cdot 100 = \$1{,}000$ |
| 200 units | $800 + 2 \cdot 200 = \$1{,}200$ |

The fixed cost is \$800, and the variable cost depends on the quantity made (e.g., at $Q = 200$ units, the variable cost is \$400). This particular formula for $TC$ can be decomposed into $FC$ and $VC$ on inspection:

$$TC = \underbrace{800}_{} + \underbrace{2Q}_{}$$

$$TC = FC + VC$$

$$FC = 800 \text{ (constant)}$$

$$VC = 2Q \text{ (varies with } Q\text{)}$$

Here's how these look on graphs versus $Q$:

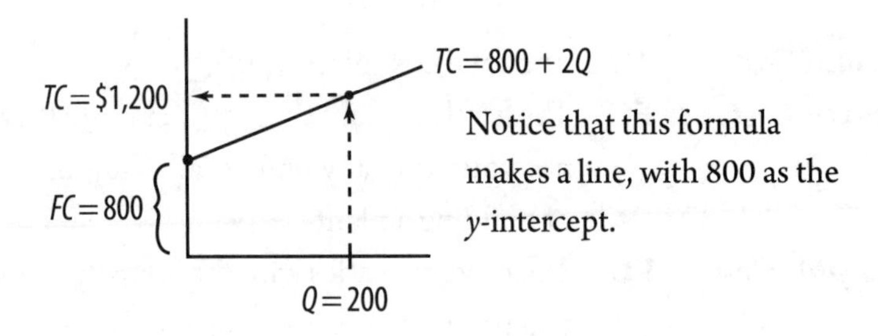

You might envision *FC* as a horizontal line stretching over from $800 on the vertical axis:

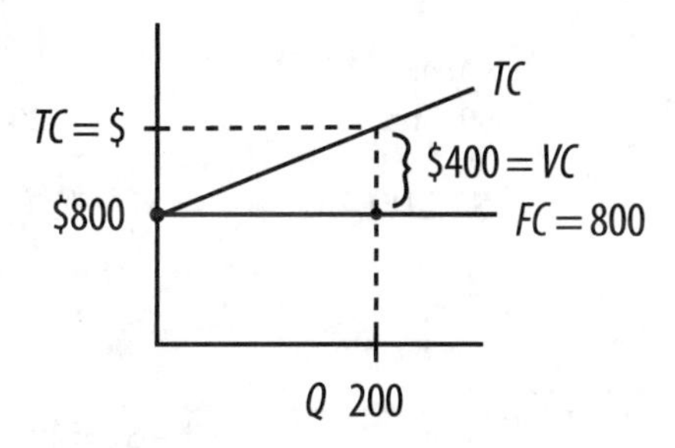


In that view, *VC* sits on top of the horizontal *FC* line. Alternatively, you might plot *VC* against *Q* as well. This way, the constant difference between the *TC* and *VC* curves is *FC*, or $800:

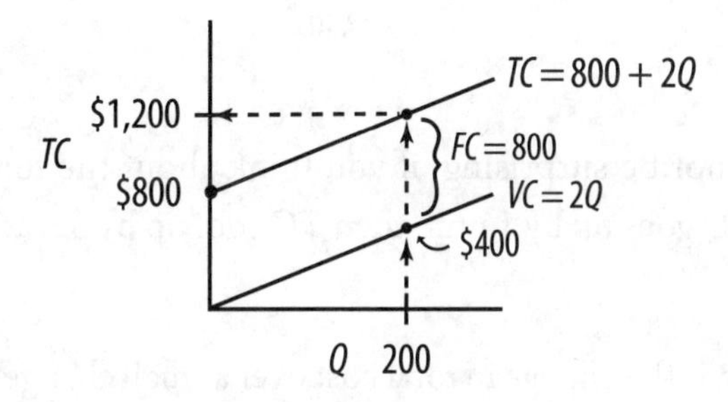


## MARGINAL COST

This concept is super-important. ***Marginal cost, MC***, is the extra cost of making *one more* unit. For instance, you already know that the cost of making 200 units is $1,200. How much would it cost to make 1 more unit? Well, first figure out how much it would cost to make 201 units:

$$TC = 800 + 2Q$$

$$\underset{\text{to make 201 units}}{TC = 800 + 2(201)}$$

$$= \$1{,}202$$

Now subtract the cost of 200 units (which we know is $1,200):

| *TC* | – | *TC* | = | *MC* |
|---|---|---|---|---|
| to make 201 units | | to make 200 units | | to make 201st unit |
| $1,202 | – | $1,200 | = | $2 (per unit) |

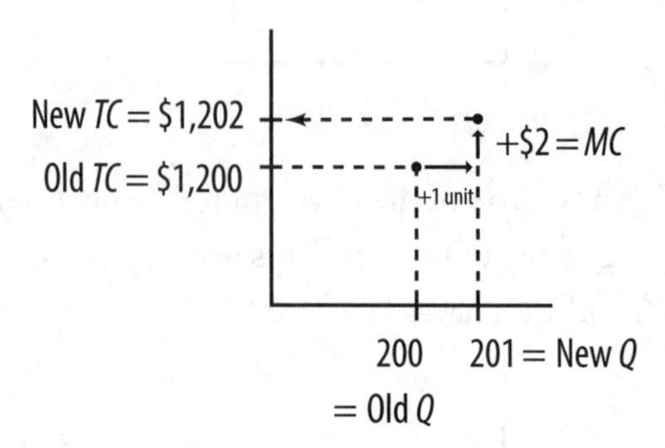


This should not be surprising, if you think about the formula $TC = 800 + 2Q$. If $Q$ goes up by 1 unit, then $TC$ goes up by $2, at any level of production.

Marginal cost is the change in total cost over a small change in quantity, usually taken to be 1 additional unit:

$$MC = \frac{\Delta TC}{\Delta Q}$$ ← usually taken to be 1 unit

Visually, *MC* is the slope of the *TC* curve when *TC* is plotted against quantity *Q*.[2] Since the formula $TC = 800 + 2Q$ represents a line with constant slope, the marginal cost is always the same in this case, meaning that it always costs $2 to make one more unit, no matter how many you've already made:

2 If you know a little calculus, you'll recognize that marginal cost is the derivative of *TC* with respect to *Q*. That is, $MC = \frac{d(TC)}{dQ}$.

+2 +2 +2 +2 $TC = 800 + 2Q$

TC

$MC = \text{slope} = \$2/\text{unit}$

Q

More generally, if the $TC$ curve is actually curved, then $MC$ will be different depending on where you are:

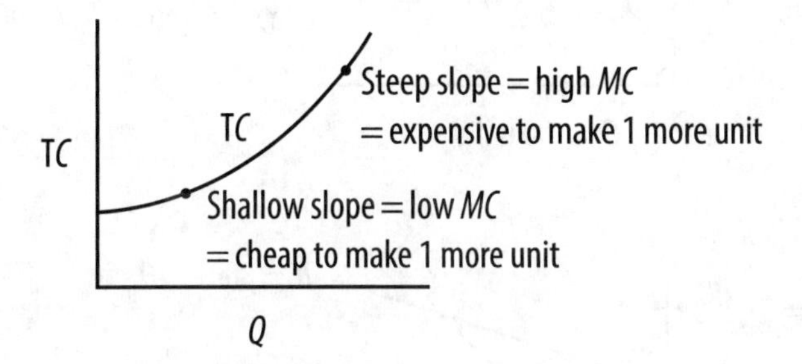


## AVERAGE COSTS

Average cost is calculated just as you'd expect. ***Average cost***, $AC$, is the total cost divided by total units produced ($Q$). For instance, if you make 100 units, then the total cost $TC = 800 + 2 \cdot 100 = \$1{,}000$, and the average cost is $\$1{,}000 \div 100 = \$10$ per unit.

$$AC = \frac{TC}{Q}$$

$$\$10/\text{unit} = \frac{\$1{,}000}{100 \text{ units}}$$

If you make 200 units, then $TC = 800 + 2 \cdot 200 = \$1{,}200$, and the average cost falls to \$6 per unit:

$$AC = \frac{TC}{Q}$$

$$\$6/\text{unit} = \frac{\$1{,}200}{200 \text{ units}}$$

Visually, average cost can be seen as the slope of a line *from the origin to the TC curve,* not the slope of the *TC* curve itself:

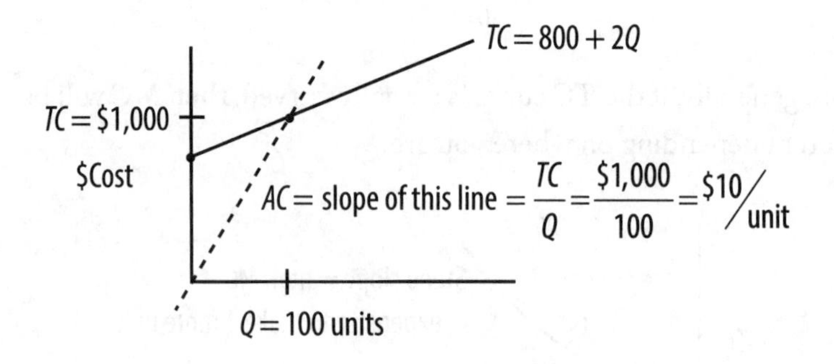


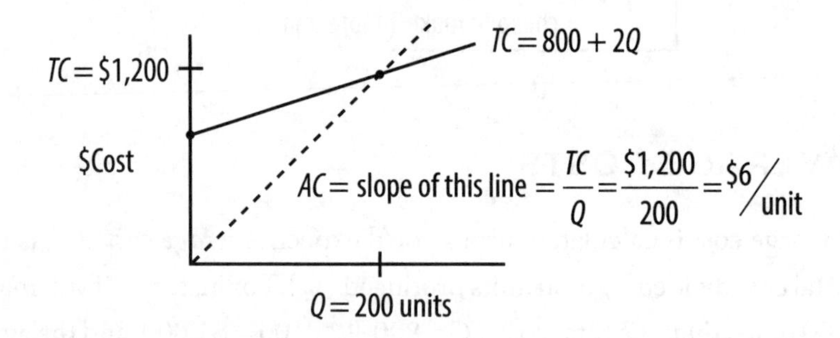


Notice the difference between marginal cost and average cost. Both are "costs per unit," but marginal cost is just the additional cost of producing a single extra unit. In contrast, average cost *includes* both fixed and variable costs, since *AC* is the total cost divided by total quantity.

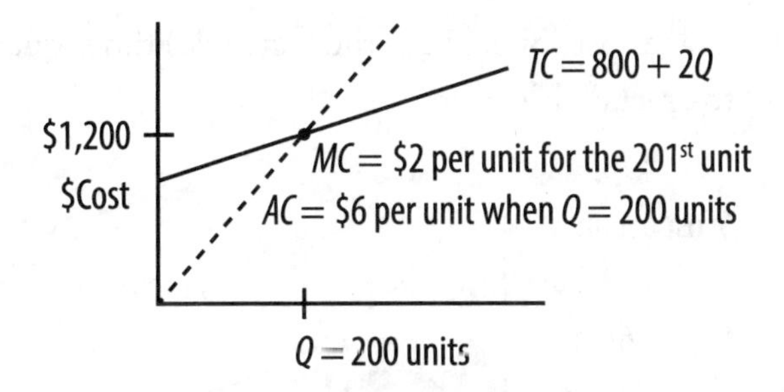


Earlier, we saw that the average cost was $10/unit at 100 units but only $6/unit at 200 units. This drop in average cost per unit can represent ***economies of scale***, which occur when you have a high upfront fixed cost (expense of building a pharma plant or writing Windows software) but a low marginal cost (expense of making one more pill or providing one more Windows download):

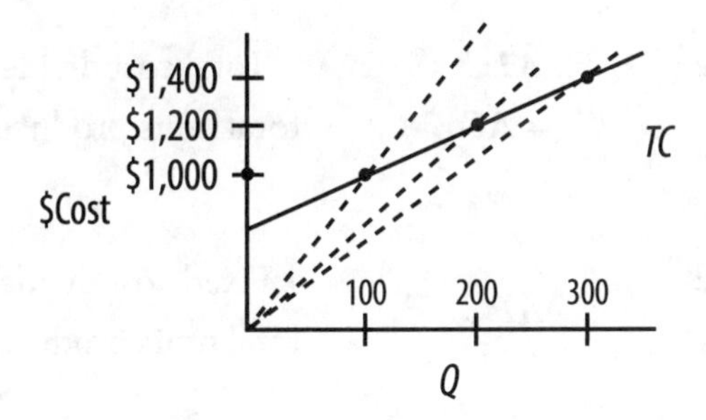


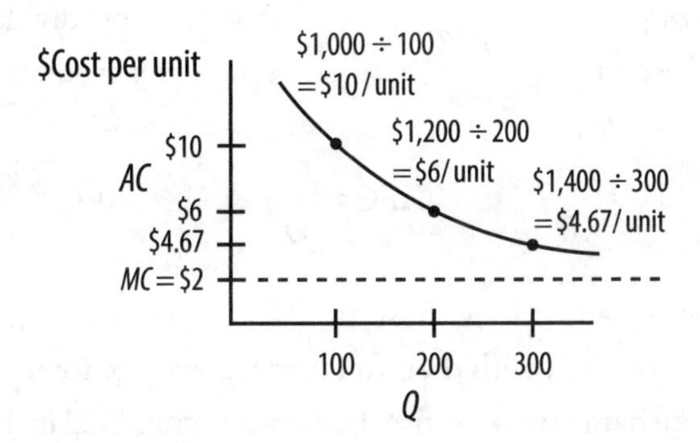


The second graph above plots costs per unit against units. On such a graph, you can think of total cost as the area of a rectangle, since total cost equals average cost (one side of the rectangle) times quantity (the other side of the rectangle):

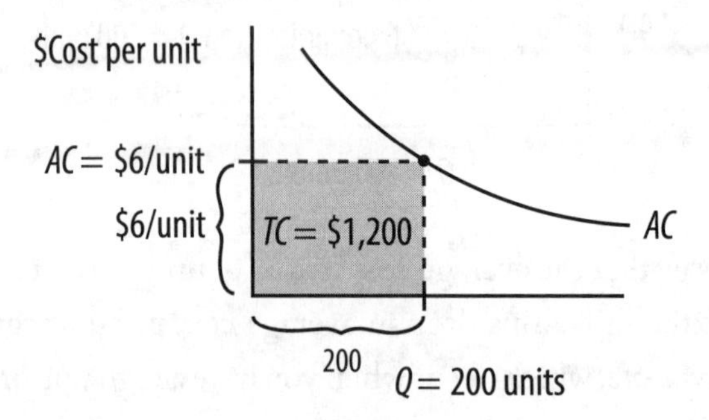


Up to now, only average *total* cost has been mentioned. You can also compute average fixed cost and average variable cost. To keep everything straight, you'll need three letters:

| | | |
|---|---|---|
| ***Average total cost*** | ***ATC* (= *AC*)** | Total cost divided by total units produced (Q) |
| ***Average fixed cost*** | ***AFC*** | Fixed cost divided by total units produced (Q) |
| ***Average variable cost*** | ***AVC*** | Variable cost divided by total units produced (Q) |

$$ATC = \frac{TC}{Q} \qquad AFC = \frac{FC}{Q} \qquad AVC = \frac{VC}{Q}$$

By the way, when you see these two- and three-letter variable names, say them to yourself with their full names: "average total cost," not "ay-tee-see." Full names make sense. Letters get scrambled up in your brain.

These three averages add up, just like their big brothers:

$$TC = FC + VC$$ ("total cost equals fixed cost plus variable cost")

Divide everything by $Q$ to take the average:

$$\frac{TC}{Q} = \frac{FC}{Q} + \frac{VC}{Q}$$

$$ATC = AFC + AVC$$ ("average total cost equals average fixed cost plus average variable cost")

For large quantities, $ATC$ often falls, because you're spreading the fixed cost over more units. Again, this phenomenon is called economies of scale (= "savings as you scale up").

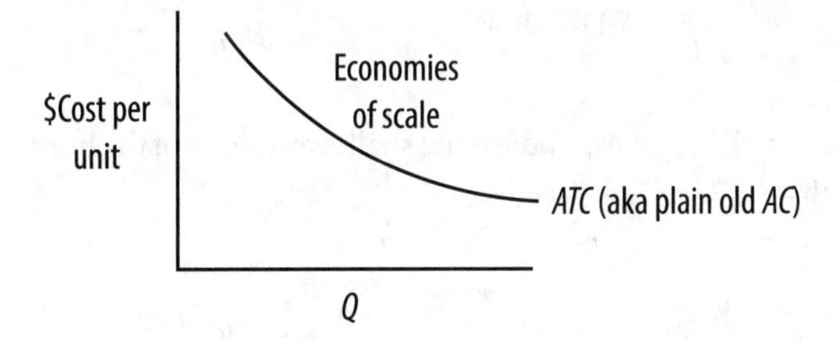


However, $ATC$ can curve *back up* if you have ***diseconomies of scale***—at some point, you have to build another plant, you have exhausted cheap supplies of raw materials, etc.

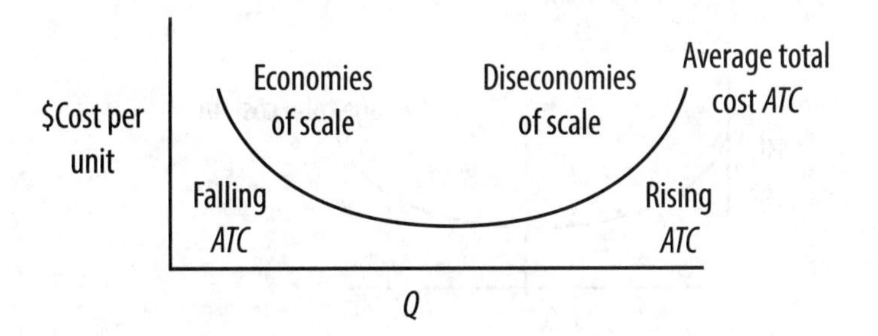

Here's what the corresponding total cost might look like:

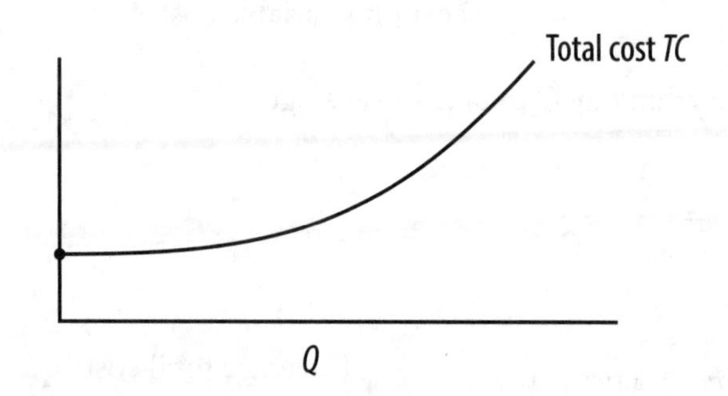


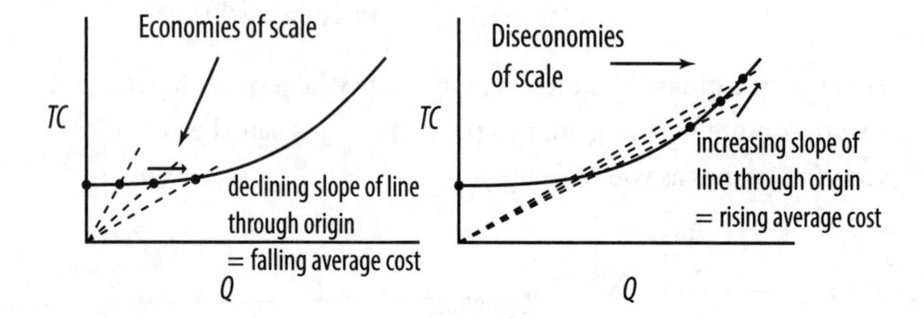


The lowest *ATC* corresponds to the shallowest slope of the line that goes from the origin to the *TC* curve:

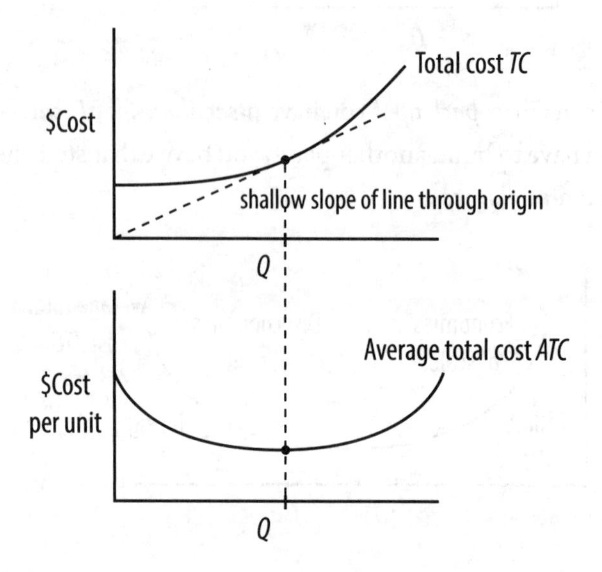

Notice that this point also corresponds to where marginal cost equals the average total cost. Here's why: remember that marginal cost tells you the slope of the *tangent* line to the total cost curve. There are three cases:

1. Economies of scale at low quantities

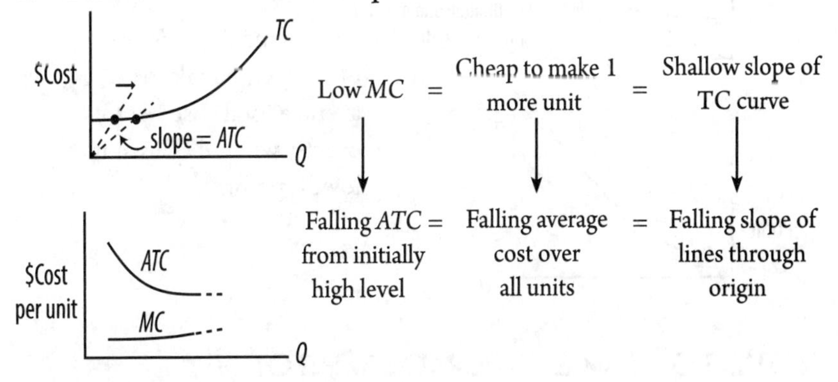


2. Diseconomies of scale at high quantities

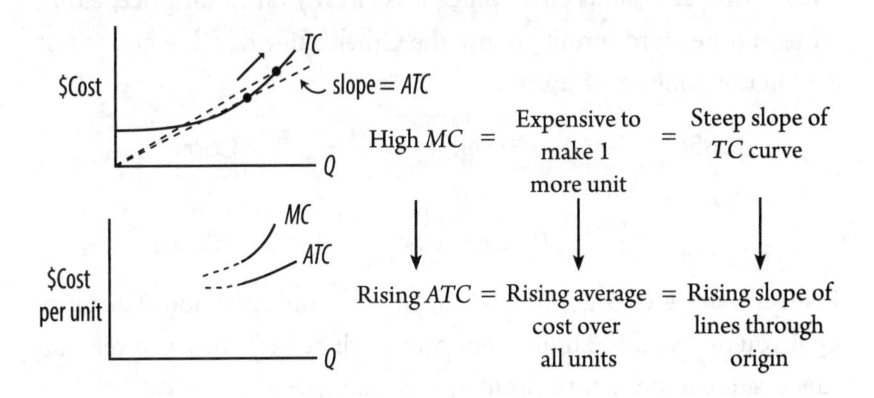

3. $MC = ATC$ (Economies are about to turn into diseconomies.)

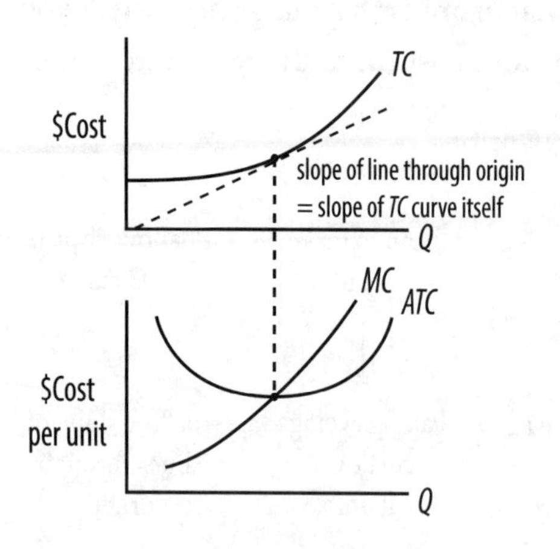


Cost of making 1 more unit ($MC$) equals average total cost ($ATC$) exactly when $ATC$ is at its lowest point.

## HOW TO MAKE THE MOST PROFIT

The last aspect of micro that there's room to talk about is profit, which equals revenues minus costs. Since $P$ is already taken for price, either write out the word "profit" or use the Greek letter $\pi$ (pi), which is not 3.14 in this context, of course.

| Profit | = | Revenues | – | Costs |
|---|---|---|---|---|
| $\pi$ | = | $TR$ | – | $TC$ |
| | | (total revenues) | | (total cost) |

Like revenues and costs, profit is thought of as some function of quantity $Q$: if you produce 100 units, your profit will be \$800 or whatever. You can imagine plotting total profit versus quantity:

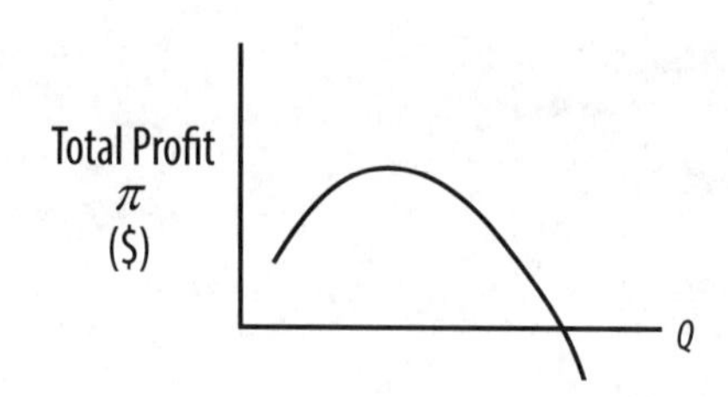

If that's your profit curve, how many units should you make and sell? Since you're a rational, self-interested profit maximizer, you'll make and sell the quantity corresponding to the biggest profit.

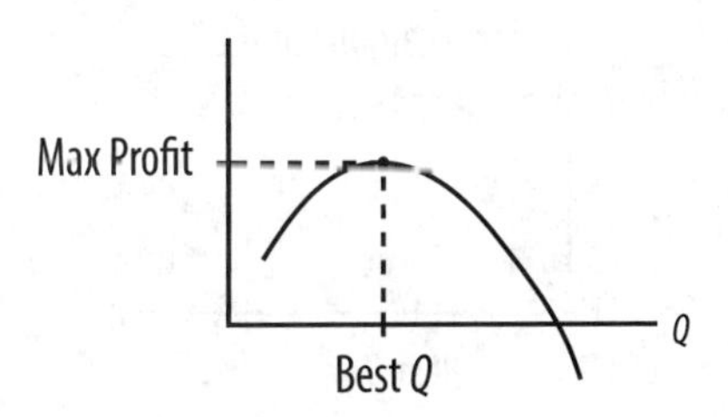


At lower quantities, you could make additional profit by producing additional units. In other words, your ***marginal profit*** (the extra profit from making and selling one more unit) is positive.

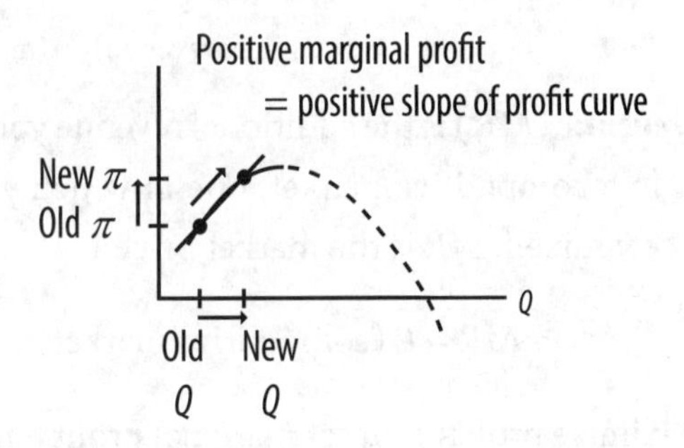


At high quantities, however, total profit is falling. Marginal profit is negative—you're actually losing money on each additional unit you make and sell.

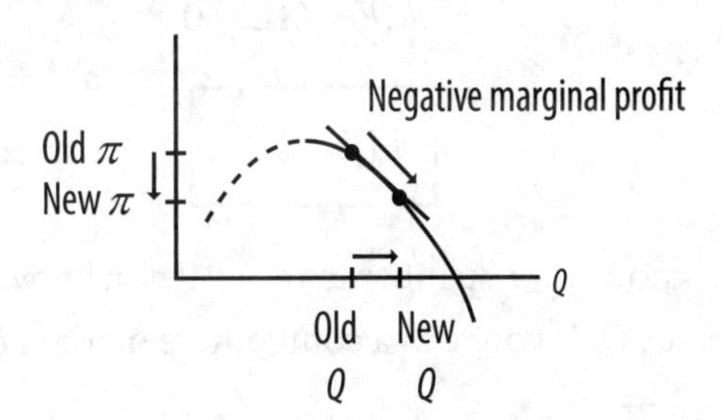

So, ***at the maximum-profit point, the marginal profit is zero***. You can't make any additional profits by making more units, but you haven't lost money on any unit yet.[3]

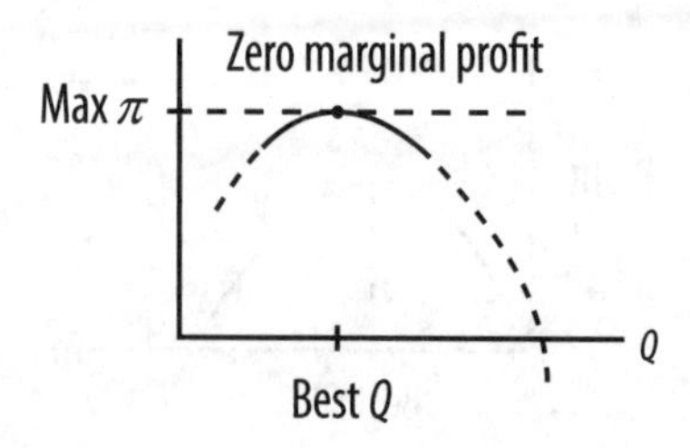


Putting this idea together with the profit equation, you get:

| | | | | |
|---|---|---|---|---|
| Profit | = | Rev | − | Cost |
| Marginal Profit | = | *MR* | − | *MC* |
| | | (marginal revenue) | | (marginal cost) |

***Marginal revenue* (*MR*)** is the additional revenue you get by selling one more unit. In a competitive market, one in which you can't affect the market price yourself, *MR* is the market price:

$$MR = P \text{ (competitive market)}$$

Now, to maximize profits, you set marginal profit equal to zero.

$$\text{Marginal profit} = 0 \text{ (to maximize profits)}$$

$$\text{Marginal revenue} - \text{Marginal cost} = 0$$

$$MR - MC = 0$$

$$\boxed{MR = MC} \quad \text{(to maximize profits)}$$

So now you know what quantity to make, if you know *MR* and *MC*. Just plot *MC* versus *Q*. If you're in a competitive market, $MR = P$, so draw a

---

3 If you've taken calculus, this idea should be familiar. The maximum of a curve shaped like this is located precisely where the slope of the curve is zero.

horizontal line corresponding to $P$. The profit-maximizing quantity is where $MC$ and $P$ cross:

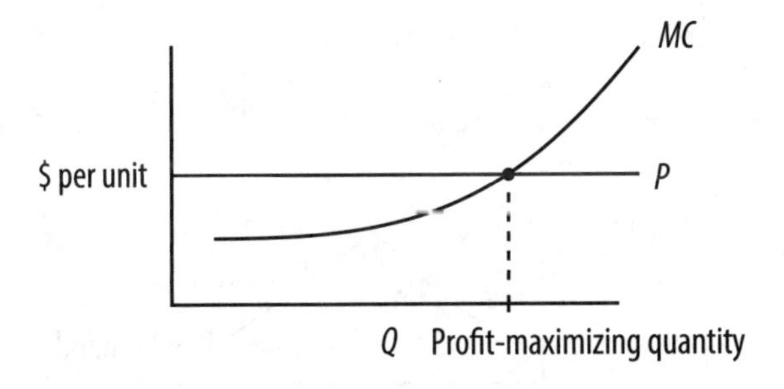


If you have a monopoly, you face the consumer demand curve completely on your own—there are no other producers. If you make more, you'll have to lower the price to unload all of your inventory, which cuts into your marginal revenue. Below, $MR = \$199 - \$100 = \$99$.

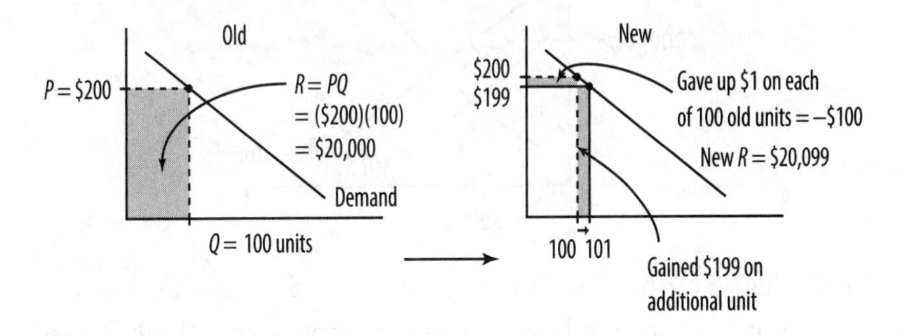


As you increase production, marginal revenue drops quickly. In fact, calculus tells us that $MR$ falls twice as fast as demand:

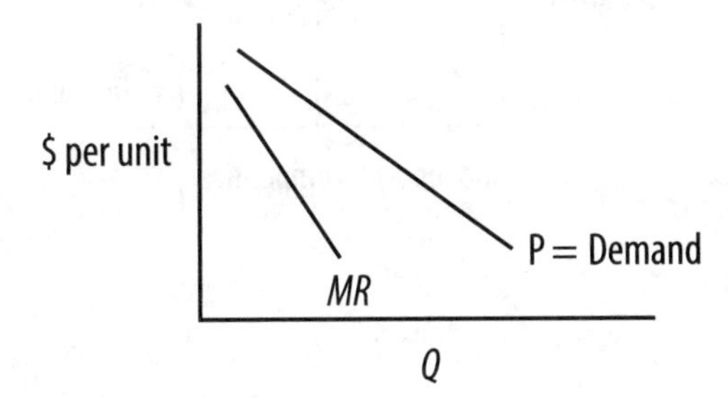

Since you're still a profit maximizer, whether you're a monopolist or not, you're going to produce the profit-maximizing quantity. This occurs where $MR = MC$.

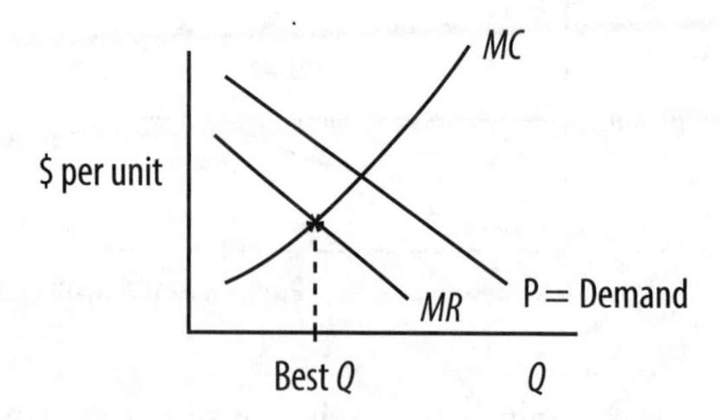


You'll wind up producing less and charging a higher price than in a competitive market:

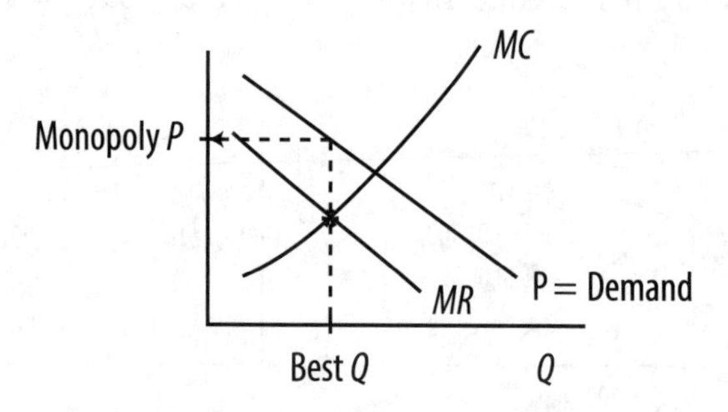


The *MC* curve can be seen as a supply curve, so under competitive conditions the market price and market-clearing quantity would be different:

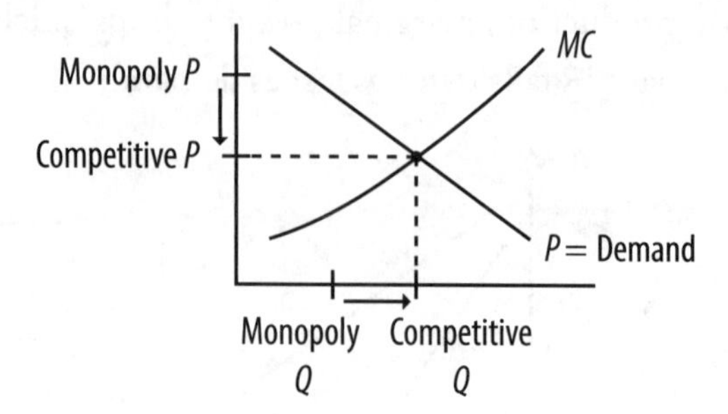

## WHAT ELSE IS THERE IN MICRO?

A couple of thoughts on remaining micro topics:

1. ***Price discrimination*** (different prices for different customers)

There are various legal and illegal ways to do this. You will make more money when you can price-discriminate, because you capture more of the ***consumer surplus*** (e.g., the $50 benefit that a customer gets when she pays $150 for something worth $200 to her). To price-discriminate legally, you can ***version*** your products (e.g., paperback vs. hardcover). You can also impose a ***two-part tariff*** (a fixed entrance fee to the park *plus* a per-ride charge). With a two-part tariff, you're effectively making different customers pay different amounts.

2. ***Trade-offs and budgeting***

Economics is all about trading off costs and benefits. This came up earlier in the "$MR = MC$" condition for profit maximization: the marginal revenue from selling one more unit should equal the marginal cost of making that unit.

The principle of balancing marginal quantities applies more generally. Imagine you had $20 to spend on Snickers and M&M's. You buy $15 of Snickers and $5 of M&M's. If that's the right split for you, you would be just as happy spending a 21st dollar (one you just found on the floor) on Snickers as you would be if you spent it on M&M's. If that's *not* the case, then you misallocated the original $20. In economic terms, at $15 on Snickers and $5 on M&M's, your ***marginal utility*** for each type of candy is equal. A 16th dollar of Snickers would make you just as much happier as a 6th dollar of M&M's would.

If you are producing Snickers and M&M's, then you do the same thing. You adjust your production mix until the marginal revenue from spending one more dollar to make Snickers equals the marginal revenue from spending one more dollar to make M&M's.

## MACROECONOMICS

Macroeconomics is concerned with the ***economies of entire countries***. "Macro" has a whole new set of concepts—and letters—but the core ideas of supply and demand developed in micro show up here as well.

Different macro schools of thought have been duking it out for years, and the current economic crisis has only intensified the debate. There are two main schools:

1. ***Neoclassical***
    - Makes the same assumptions as seen earlier in microeconomics
        - Firms and individuals are rational, self-interested, and profit-maximizing.
        - Competition is perfect and information flows freely.
    - Trusts ***free markets*** (the market mechanism unhindered by taxes, etc.) to do the best job of allocating resources efficiently and matching supply and demand
        - Market for labor
        - Market for capital (wealth to invest)
        - Market for money itself
2. ***Keynesian*** (pronounced "kane-sian"), after economist John Maynard Keynes
    - Agrees with neoclassical economics in most respects.
    - Argues that ***wages and prices are sticky***.
        - Wages adjust slowly, so the labor market takes a "long time" to reach equilibrium (how long exactly is open to debate).
        - The government has a role in making up for this private-sector lag.
        - When private demand is low, the government should stimulate demand by spending money and by creating money.

- When private demand is high, the government should restrain demand by taking in money (through taxes) and by destroying money.
- In this way, according to Keynes, the government's ***fiscal policy*** (taxation and spending) and ***monetary policy*** (money creation and destruction) help smooth out the ***business cycle***—the ups and downs of economic activity observed in the world.

Although these two schools bitterly diagree about the proper role of government in the economy, they share many features. In b-school, you'll learn the neoclassical model primarily, since Keynes can be seen as a variation. Then you'll try to argue with your professors; good luck with that.

## CIRCULAR FLOW

*Macro* means "big," literally, so from the biggest point of view, here's what happens in a simplified national economy (one with no taxes, no savings or investment, and no outside trade):

1. People work for companies.
2. People buy stuff from companies.

That's it! When people work for companies, they provide labor and get paid wages:

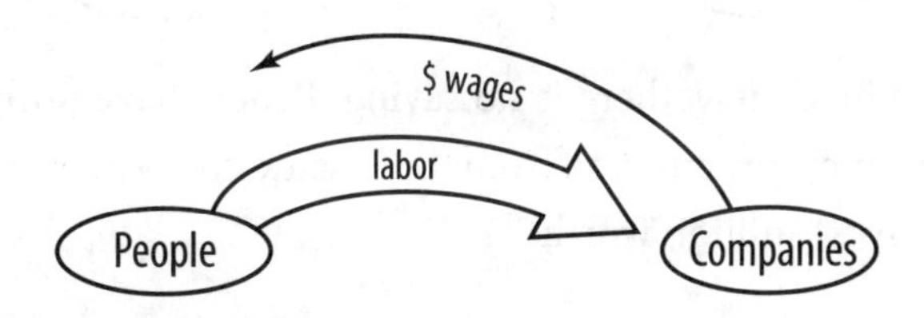


People can also sell materials or rent out their land or equipment. They are providing ***factors of production*** to companies, in return for money. As owners of companies, people can receive profits as well.

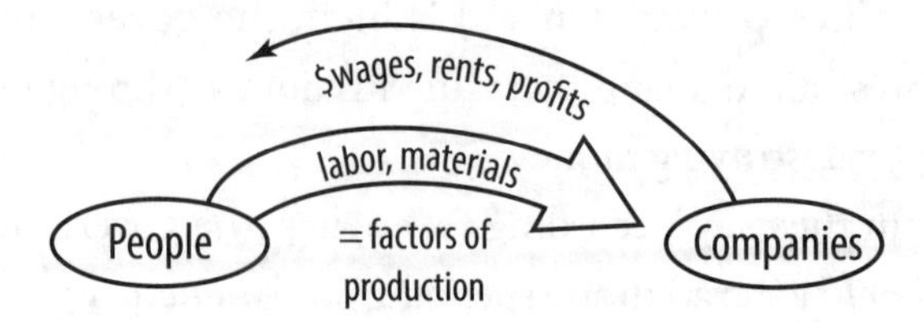


In turn, people buy goods and services from companies.

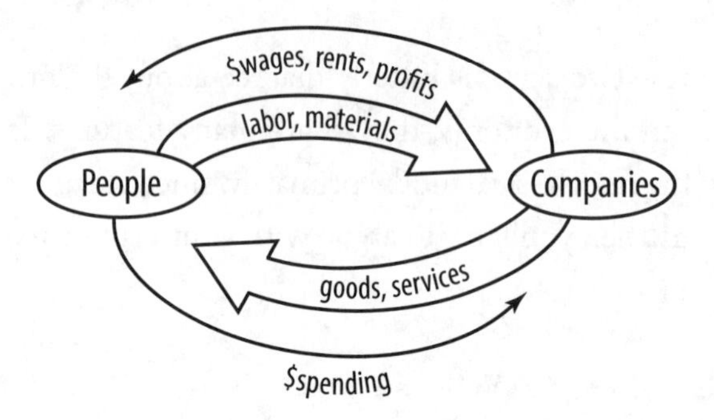


This way, there's a complete cycle of trade. Valuable stuff makes a full circle, and money also makes a full circle in the opposite direction. Imagine that over the course of a year, in a *very* tiny economy, people earn 1 million dollars for their work.

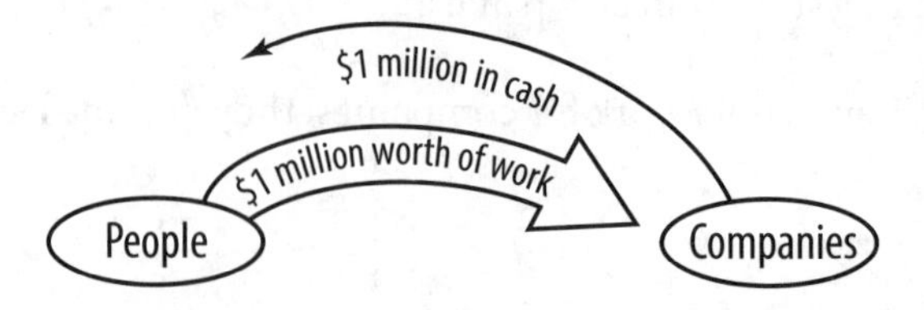


In this simplified view, there is no saving. People have nothing to do with their money but buy stuff from those same companies. How much stuff? Exactly $1 million worth.

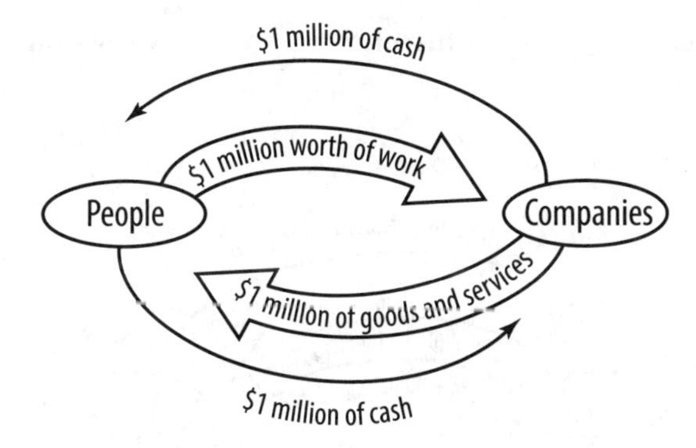


Notice that every flow on the diagram is worth $1 million. This $1 million of "economic activity" can be looked at in several different ways:

1. ***Production***: the total value of goods and services produced. This is called ***gross domestic product***, or **GDP**, or ***output***.

   Note that you exclude the value of any *intermediate* products bought and sold between companies. Alternatively, you can count them, but then you subtract their value from the *final* goods consumed by people so that you're just counting ***value added*** at every step. Either way, you get $1 million.

2. ***Income***: the money earned by people, whether as wages, rents, or profits. This is called ***national income***, represented by the capital letter $Y$ (don't ask us why).

3. ***Consumption***: the money spent by people. This is given the letter $C$.

In this super-simplified economy with just people and companies (or households and firms, to use fancy names), there's a three-way ***national income equation***.

| Production | = | Income | = | Consumption |
|---|---|---|---|---|
| *GDP* | = | *Y* | = | *C* |

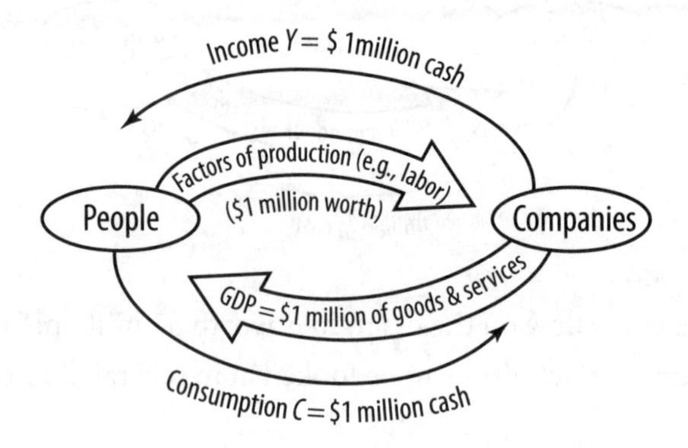


## OTHER PLAYERS

To make the model more realistic, additional players can be added. These players will split the flows, but it'll all still form a circle.

### *Government*

The government takes in taxes $T$ and spends that money on goods and services. For simplicity, imagine that the government cannot run a ***budget deficit*** or a ***budget surplus***—it must spend exactly what it takes in in taxes, no more and no less—and that the government imposes a 25% income tax on people only and provides an equal amount of benefits in return.

In a sense, all that happens is that the government gets inserted into the "bottom" flows on the simple picture above, as 25% of the money spent by consumers and 25% of the goods and services produced by companies are diverted through government hands.

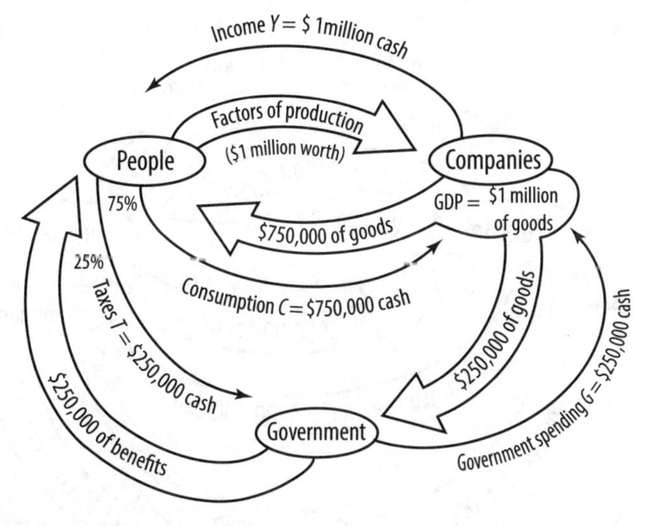


The adjusted equation is this:

| Production | = | Income | = | Expenditures | | |
|---|---|---|---|---|---|---|
| $GDP$ | = | $Y$ | = | $C$ | + | $G$ |
| | | | | personal consumption | | government spending |
| $1 million | = | $1 million | = | $750,000 | + | $250,000 |

## FINANCIAL SECTOR

Putting in a financial sector allows for savings and investment. If you assume that only people save 10% of their income and that the financial sector invests every dollar of savings, the picture looks similar to the previous one, with the financial sector in place of the government. That is, a fraction of the bottom flows of both money and goods/services move through the financial sector. In this case, these diverted flows are not taxes and government spending, but rather savings and investment.

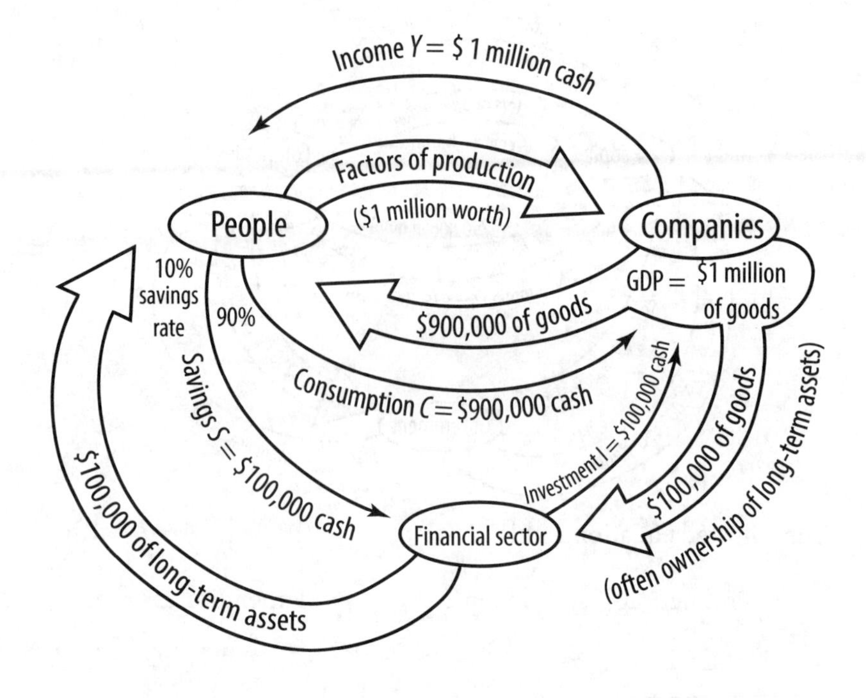


The national income equation adjusted for the financial sector is this:

| Production | = | Income | = | Expenditures | | |
|---|---|---|---|---|---|---|
| $GDP$ | = | $Y$ | = | $C$ | + | $I$ |
| | | | | personal consumption | | investment |
| $1 million | = | $1 million | = | $900,000 | + | $100,000 |

The national income equation adjusted for *both* the financial sector and the government is this:

| Production | = | Income | = | Expenditures | | | | |
|---|---|---|---|---|---|---|---|---|
| $GDP$ | = | $Y$ | = | $C$ | + | $G$ | + | $I$ |
| | | | | personal consumption | | government spending | | investment |

Note that $I$ stands for investment, not income (which is $Y$).

***Other Countries***

Up to now, the picture has been of a ***closed economy*** with no international trade. An ***open economy*** interacts with those of other countries:

$X$ = ***exports*** = currently produced goods you sell to other countries
= the money they pay you for those goods

$M$ = ***imports*** = currently produced goods other countries sell to you
= the money you pay them for those goods

$NX$ = ***net exports*** = $X - M$

So if you export \$300,000 of goods and import \$100,000 of goods, your $NX$ = \$300,000 − \$100,000 = \$200,000. Here's the start of the picture with other countries added in:

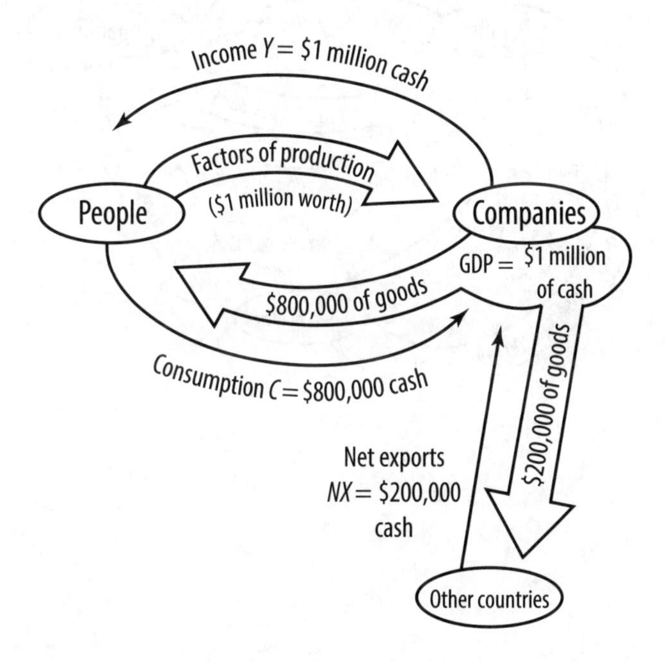

Notice that the flow is unbalanced. Where did those other countries get the $200,000 in cash to buy the exported goods? If you look around the diagram, you see that "People" on the left are earning $1 million but only spending $800,000 as current consumption. So your people (in your country) are $200,000 richer. To complete the flow, they give that $200,000 cash to other countries in exchange for *something*—but not currently produced goods or services, since those would be counted as *imports*, which are already included in the *NX* number of $200,000. In other words, *after* subtracting out imports, there's still a net outflow of goods to other countries; those goods must be paid for. (Pretend there is no borrowing or currency reserves lying around.) The only way for the other countries to get the extra $200,000 in cash to pay for the net exports is to sell *long-term assets* to your people, who are frankly saving that money—socking it away for the long term.

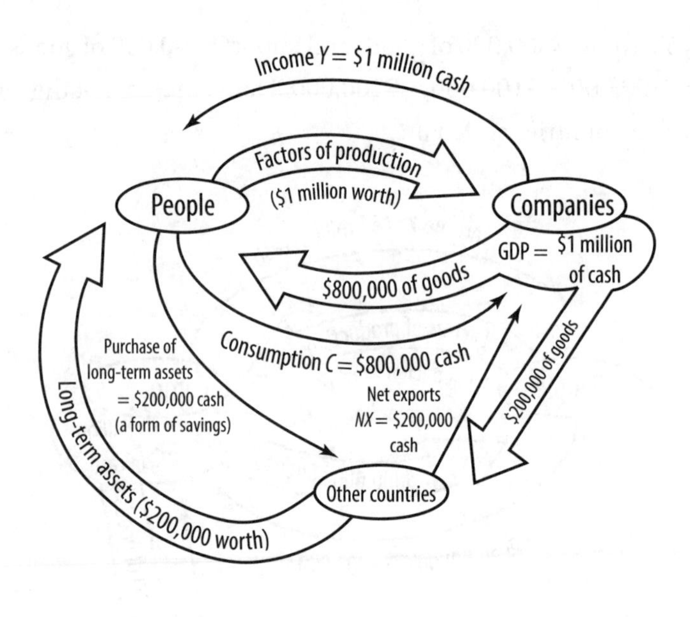

The national income equation adjusted for net exports is this:

| Production | = | Income | = | Expenditures | | |
|---|---|---|---|---|---|---|
| $GDP$ | = | $Y$ | = | $C$ | + | $NX$ |
| | | | | personal consumption | | net exports |
| $1 million | = | $1 million | = | $800,000 | + | $200,000 |

Putting all the expenditures together, you get the big kahuna equation:

| Production | = | Income | = | Expenditures | | | | | | |
|---|---|---|---|---|---|---|---|---|---|---|
| $GDP$ | = | $Y$ | = | $C$ | + | $I$ | + | $G$ | + | $NX$ |
| | | | | personal consumption | | investment | | government spending | | net exports |

## LETTER ZOO

The capital letters don't stop there—macro is thoroughly infected with them. A complete rundown would take too many pages, but there are two more relationships you should know:

1. $CA + KA = 0$

This equation has to do with international trade. The ***current account*** $CA$ tracks payments you receive from other countries for currently produced goods and services (minus what you pay other countries for *their* current goods). $CA$ includes net exports $NX$, as well as ***net factor payments*** $NFP$ (money that your people and companies are paid for work they do outside your country) and any gifts or foreign aid you receive. In the last diagram, you received $200,000 in $NX$, so your $CA$ was +$200,000.

In contrast, the ***capital account*** $KA$ tracks payments you receive as capital investments, in exchange for long-term assets (minus what you pay other countries for their long-term assets). In the diagram, you paid $200,000 out to other countries for their long-term assets, so $KA = -\$200{,}000$. Notice that $CA + KA = 0$. This is always true.

By the way, $K$ is used for "capital" in macro. Maybe this is the legacy of Karl Marx (*Das Kapital*)?

2. $S_{private} = I + (-S_{gov't}) + CA$

This equation lists the three ***uses of private saving*** $S_{\text{private}}$ (savings done by people or companies, not the government):

a. Provide investment $I$ to companies, either directly or through financial institutions
b. Offset government borrowing, which is negative saving $(-S_{\text{gov't}})$
c. Allow foreigners to purchase your current products ($CA$), because you're buying long-term assets from those foreigners

## THE PRODUCTION FUNCTION

National income $Y$, also known as output, can be thought of as a *mathematical* output too. You feed it two inputs, and *voilà*—you have $Y$. The two inputs are ***capital*** $K$ and ***labor*** $N$, where $N$ = number of workers (some books use $L$ for labor, but $L$ shows up elsewhere).

Capital $K$, by the way, is generally the capital *stock* in the country: the machines, buildings, and others productive assets that help generate economic activity.

The magic transformation into $Y$ can be written this way:

$$Y = A \cdot F(K, N)$$

where $A$ is a constant known as ***total factor productivity,*** or *TFP*. Remember this number—it's like crack to macroeconomists. If you want to make an economy better, you've got to raise *TFP*. That is, you've got to make each worker and each dollar of capital turn more *effectively* into dollars of output $Y$. This is the key to building national wealth over the long term.

Meanwhile, $F(K, N)$, which you can call "$F$ of $K$ and $N$," is just some mathematical expression involving $K$ and $N$. The key is that it should show ***diminishing returns***—that is, another thousand workers should boost national income a lot more when you only have a few workers to begin with than when you already have zillions. The same holds true for capital investment: investing the first billion dollars does more to raise national income than the next billion, which does more than the next, and so on.

Thus, the relationship between $Y$ and $K$ or between $Y$ and $N$ should look like this:

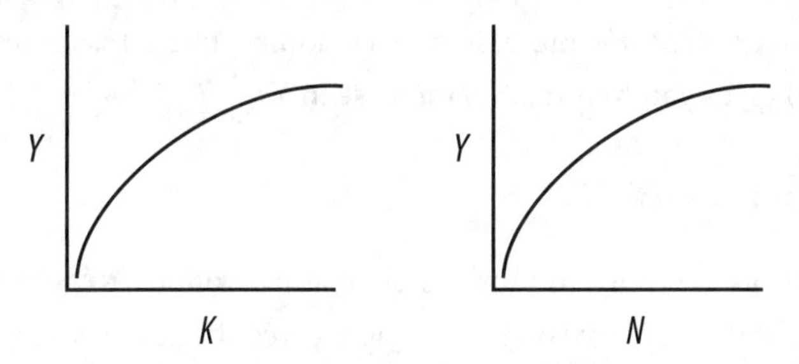


Flatter slopes out to the right mean diminishing returns. Here's a possible form of the production function:

$$Y = A \cdot K^{0.3} \cdot N^{0.7}$$

That's $K$ to the 0.3 power and $N$ to the 0.7 power. What do decimal exponents even mean? Well, $x^{0.5}$ is $x^{½}$, which is another way to write $\sqrt{x}$. The square root of $x$ looks like this on a graph:

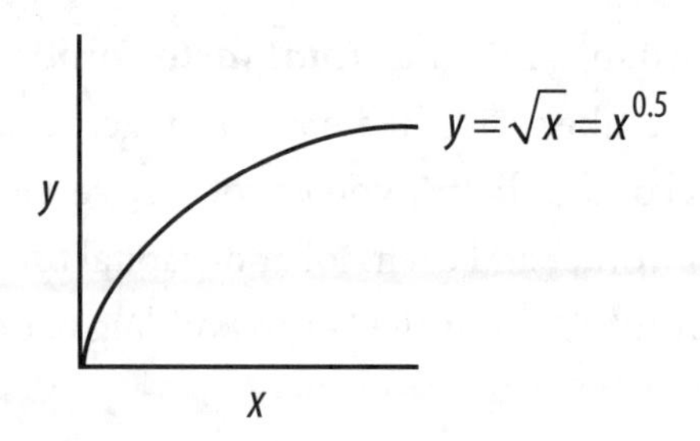


So those decimal exponents, like 0.3 on $K^{0.3}$, just create the diminishing returns we want.

The production function can also be written in terms of percent changes. You can derive this second form using calculus, or you can just remember that when you take percent changes, multiplication turns into addition, and exponents turn into multiplied coefficients.

1. $Y = A \cdot K^{0.3} \cdot N^{0.7}$
2. $\%\Delta Y = \%\Delta A + (0.3)(\%\Delta K) + (0.7)(\Delta\% N)$

This second form tells you that a 1% increase in $A$ (total factor productivity) gives you a 1% increase in $Y$ (income), but a 1% increase in $K$ (capital) gives you only a 0.3% increase in $Y$.

## MONEY SUPPLY

There is only a certain amount of cash floating around; as Mom and Dad told you, it doesn't grow on trees. Macro cares about the ***money supply***, because a country's economy can go haywire if adults aren't paying attention to how much cash is out there.

The most obvious form of money is physical ***currency*** (bills and coins), but many other things are "so money" (or at least "more or less money") because they have these characteristics.

- Measurable value
- Stable value
- Accepted in exchange for other assets } high ***liquidity***
- Convertible to currency

The money supply is referred to with even more letters and numbers. ***M1*** is the most liquid money aggregate and most narrowly defined. It includes not only the cash in your pocket but the cash in your checking account (***demand deposits***). ***M2*** is a broader aggregate, including all of *M*1 *and* the cash in your savings account and money market accounts. There are even broader aggregates, if you care to know. When you don't care exactly which money definition you're using, you just write plain old *M*.

## PLAYING GOD WITH MONEY

The ***Federal Reserve*** is the central bank of the United States. It has the authority to print money—and to destroy money if it's being bad. When the Fed prints money, it doesn't airdrop Benjamins over Central Park (although one could argue that'd be pretty effective for some of the Fed's ultimate purposes). Rather, the Fed prints money and uses it to buy government bonds in the market. That's how the cash gets into circulation. The Fed doesn't always literally hand over bricks of printed greenbacks, but in theory that's what it's doing.

Moreover, a small purchase of government bonds can turn into a lot of circulating money through the ***money multiplier*** effect.

The Fed prints $100 and buys a government bond from Bank *X*. Bank *X* now is sitting on $100. Since banks are *theoretically* in the business of lending money, Bank *X* can now lend out this $100, or most of it anyway, keeping behind a certain percentage to comply with the Fed's ***reserve requirement***. Say the bank has to keep 10% on hand, or $10 in this case. Then Bank *X* lends $90 to Company *Y*, which turns around and spends it on computers from Company *Z*. Company *Z* deposits the $90 with its bank, Bank *W*, which then turns around and lends $81 out...

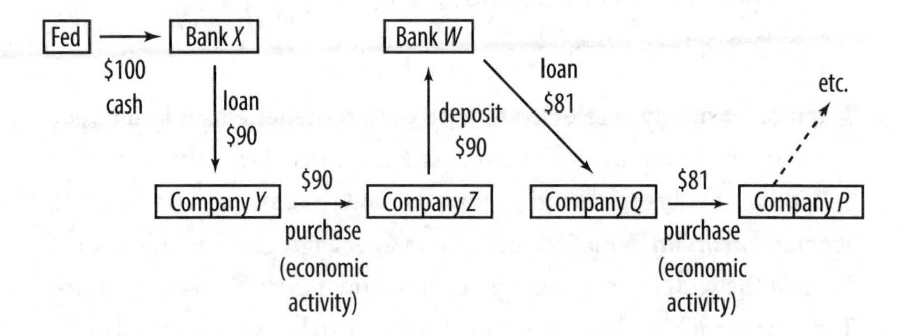


If banks are lending and companies aren't just sitting on their cash, that first $100 can turn into many hundreds of dollars of economic activity. This is how the Fed's $100 is multiplied through the economy.

Likewise, if the Fed *sells* a government bond to the public, money is pulled *out* of circulation, with a larger ripple effect at some multiple of the purchase.

## INFLATION

Your grandma kvetches that things cost more now than they did 50 years ago. She's right, sort of. ***Inflation*** is the rise in the ***general price level*** *P*. (Note: *P* here doesn't mean the price of any specific thing, as it did in micro. In macro, *P* tracks the overall rise in prices across the board.) *P* can be measured by something like the ***consumer price index, CPI*** (also known as the consumer ***deflator***, which sounds like a rather specialized superhero, perhaps one who lets the air out of tires on getaway cars).

So if the *GDP* of your little country is $1 million this year, and it was $900,000 last year, how much of that roughly 11% growth was *real* growth in the amount of goods and services your country makes—and how much was just inflation, just a rise in the overall prices of things?

Well, you've got to take these ***nominal*** dollar figures (the \$900,000 last year and the \$1 million this year), which represent the physical dollars moved around each year, and put them in ***real*** terms. That is, you choose a reference year—say, 2018—and put everything in terms of dollars from that year.

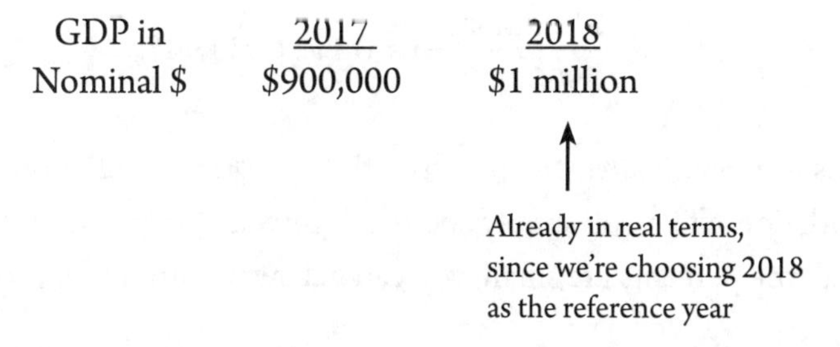


How do you turn \$900,000 of 2017 dollars into 2018 dollars? You increase that amount by the inflation rate. Say inflation was 3% from 2017 to 2018. Then you grow the \$900,000 by 3%.

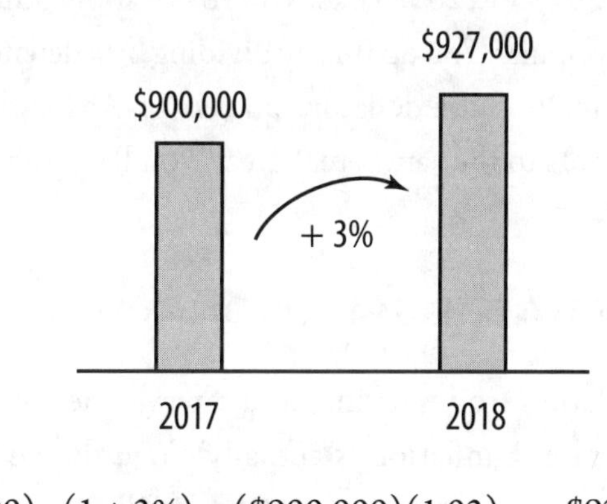


$$\underbrace{(\$900{,}000)}_{\text{in 2017 dollars}} \cdot \underbrace{(1+3\%)}_{\text{inflation rate}} = (\$900{,}000)(1.03) = \underbrace{\$927{,}000}_{\text{in 2018 dollars}}$$

So now you can compare the *GDPs* in real terms, apples to apples:

| | 2017 | 2018 |
|---|---|---|
| *GDP* in Real \$ (2018 dollars) | \$927,000 | \$1 million |

The percent growth of the *GDP* in real terms is

$$\frac{\$1\text{ million}}{\$927{,}000}-1\approx 0.0787=7.87\%.$$[4]

By the way, the percent growth in nominal terms was

$$\frac{\$1\text{ million}}{\$900{,}000}-1\approx 0.1111=11.11\%.$$

This is *approximately* the real growth (7.87%) plus inflation (3%). The math doesn't let you simply add these percentages precisely, but as long as all the percents are small, you can get away with this approximation:

$$\underset{(\%)}{\text{Real Growth}} + \underset{(\%)}{\text{Inflation}} \approx \underset{(\%)}{\text{Nominal Growth}}$$

Often, the reference year is somewhere in the past, and you bring both modern figures (2017 and 2018) back to 1987 or some other year when Huey Lewis was popular. You do this by dividing by a deflator calibrated to 1987 (that is, in 1987 the deflator equaled 1). As long as you bring all nominal amounts to the same "real" year, you'll get the same results in the end.

## INFLATION BAD, DEFLATION BAD

A high rate of inflation erodes savings and can even destroy economies. However, a low rate of inflation is actually okay: the Fed's unofficial target rate is 2%, not 0%. In fact, ***deflation*** (a fall in the general price level) is really bad. You'd think it'd be fine, until you realize that it freezes economic activity. Cash becomes *more* valuable over time, so you don't spend it. Banks stop lending, businesses stop investing. For a little while, the falling prices seem great to you as a consumer, until your own wages get cut because your company's products are sold for less and less. Ugly.

---

4 This just uses the percent change formula: Percent Change $=\frac{\text{New Amount}}{\text{Old Amount}}-1$, then express the decimal as a percent.

Let's avoid this over the next number of years—there's a real risk of deflation right now. So go out and spend wildly in da club. By purchasing bottles full o' bub, you're helping prevent a deflationary spiral.

## MONEY DEMAND

The term ***money demand*** might seem obvious—everybody wants money. That's not what the macro concept of money demand is about. Rather, it's this: given that people and companies have, say, $1 billion in wealth, how much do they want to hold as money (currency, demand deposits) rather than in other nonmonetary forms? Money's liquid (exchangeable), but it earns no interest (or very little—your checking account might pay a trivial rate), so there's a trade-off between holding money and holding ***nonmonetary assets*** (stocks, bonds, real estate, etc.). Say that, out of $1 billion in total wealth, people and companies want to keep $100 million in the form of liquid money. You can write:

$$M^d = \$100 \text{ million}$$

The little *d* stands for "demand."[5]

$M^d$ is *nominal* money demand. To express this demand in *real* terms (factoring out inflation), you can write:

$$\text{Real money demand} = \frac{M^d}{P}$$

where *P* is the general price level, as mentioned above. Real money demand is represented by the capital letter *L*, for *liquidity*. Two factors influence *L*.

5 It's not an exponent—macroeconomists should use subscripts instead and write $M_d$ instead of $M^d$, but even economists don't always act rationally.

1. National income $Y$

The higher $Y$ is, the more money is needed for transactions throughout the country. So money demand $L$ goes up. This should make sense at an individual level too. Rich cats with a high income are the ones carrying the fat rolls of Benjamins.

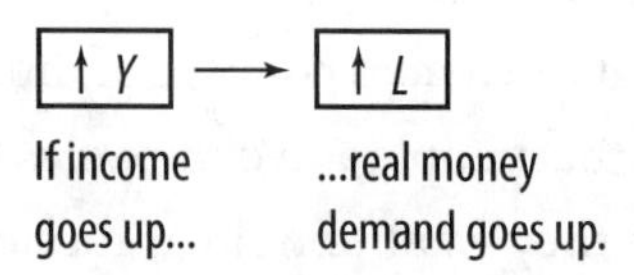


2. ***Real interest rate r*** on nonmonetary assets

Remember, people and companies have a choice about where they stash their wealth.

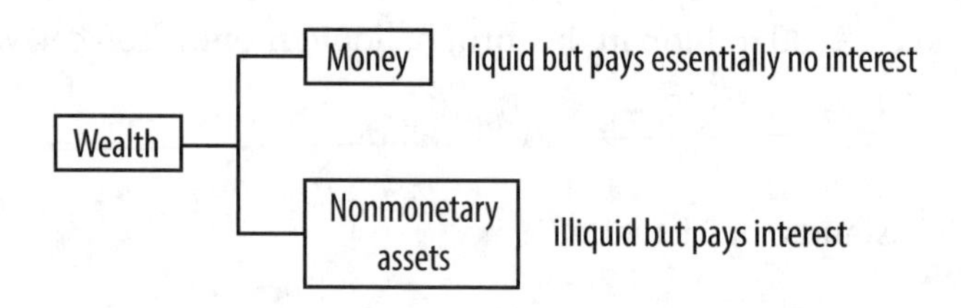


So if you keep \$1,000 cash in your pocket at all times, you're choosing to give up the interest $r$ you could earn on that \$1,000 if you invested it in some nonmonetary asset. Thus, you can think of $r$ as the price of money—that's what it costs you to keep the \$1,000 as liquid cash, earning zilch for interest.

As $r$ goes up, the price of holding money (instead of investing it at $r$) goes up, so the demand for money ($L$) goes down. This is a typical demand curve.

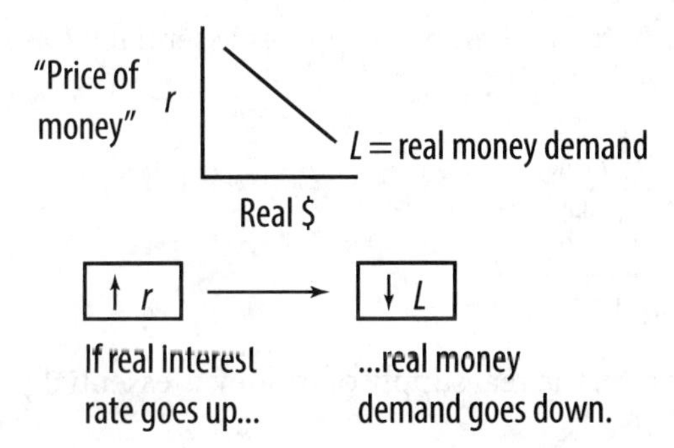


If you can invest wealth at 20%, you're more apt to stick the $1,000 in the 20%-paying investment (which is certainly not a monetary asset), because it's costing you $200 a year to carry around $1,000 in cash. On the other hand, if $r =$ only 1%, it's only costing you $10 a year to carry $1,000 in cash, so you're likely to keep more of your assets in cash or other monetary forms. An increase in the expected rate of inflation also decreases money demand, since nominal interest rates would be higher as well.

## MONEY SUPPLY AND MONEY DEMAND

If real money demand looks like this—

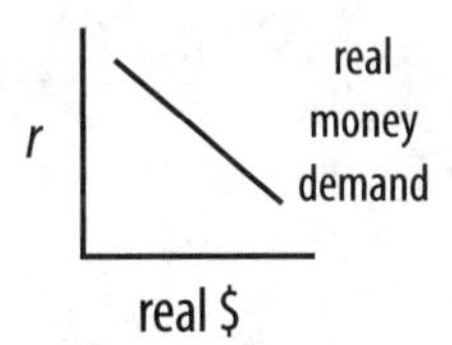


does ***real money supply (MS)*** look like this?

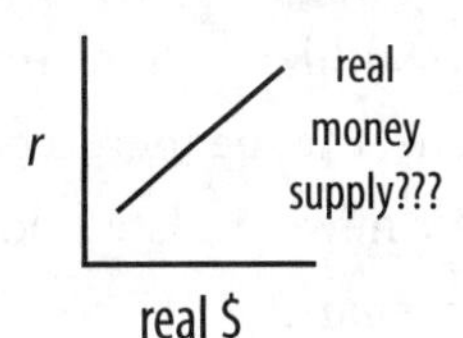

The answer is no: the real money supply is essentially *fixed*:

r

real money supply *MS*

real $

At any interest rate, the real supply of money is essentially constant.

What about all that stuff earlier about the Fed creating and destroying money? That was all *nominal* money. In the short run, the Fed may be able to goose or stifle economic activity by creating or destroying nominal dollars, but people aren't stupid. Take an extreme example. One morning, to really jump-start the economy, the Fed makes a crazy announcement:

Ben Bernanke
Fed chairman

Yes, this is insane, but what would theoretically happen? Every business would add a zero to every price. The general price level *P* would jump by a factor of 10, but no one would *really* be wealthier, and the real money supply, for one thing, would remain the same. Maybe there'd be a short-term spurt as everyone tries to spend their newly inflated dollars, but prices would adjust pretty quickly in this scenario.

$$\frac{M}{P} = \text{Real money supply } MS \approx \text{constant}$$

(Nominal money supply → $M$; General price level → $P$)

What neoclassical economists and Keynesian economists disagree about, among other things, is *how quickly* prices would adjust.

Neoclassical: *P* adjusts instantly.

Keynesian: *P* takes time to adjust.

But both sides agree that, in the long run, *P* adjusts to cancel out fluctuation in the nominal supply of money *M*.

So now you can put real money supply and real money demand on the same graph.

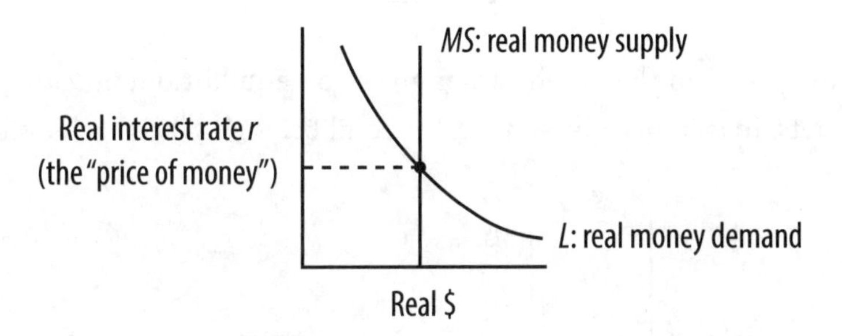


Where the two curves intersect determines the real interest rate in the economy.

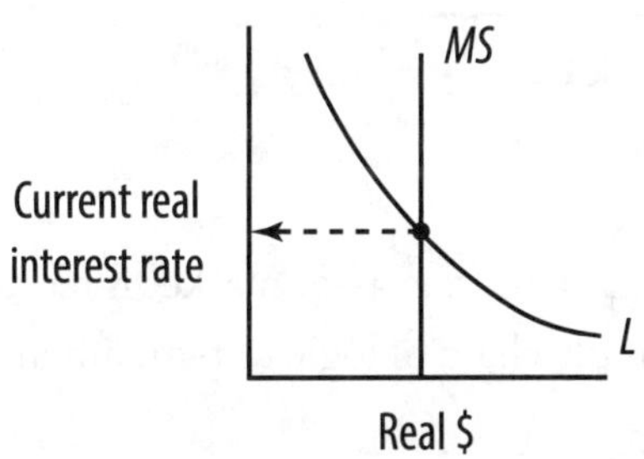


What's cool about this point is that it does two jobs at once. Most directly, it tells you where money supply equals money demand. It tells you that:

$$\text{Monetary assets supplied} = \text{Monetary assets demanded}$$

at some magical interest rate *r*.

But you get two for one. Every choice about where to keep your wealth is a choice between monetary and nonmonetary assets. (In fact, that "price of money" *r* is the interest you could earn on nonmonetary assets.) So if your economy matches up supply and demand on the *money* side, then it also matches up supply and demand on the *non-money* side.

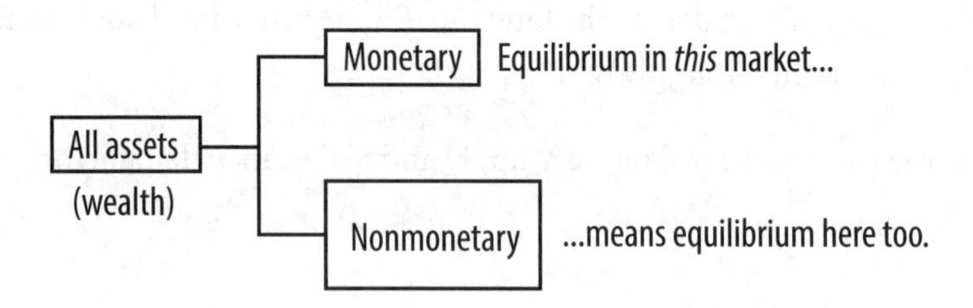


So the point on the graph earlier gives you equilibrium in *both* asset markets. In fact, overall asset market equilibrium has been achieved.

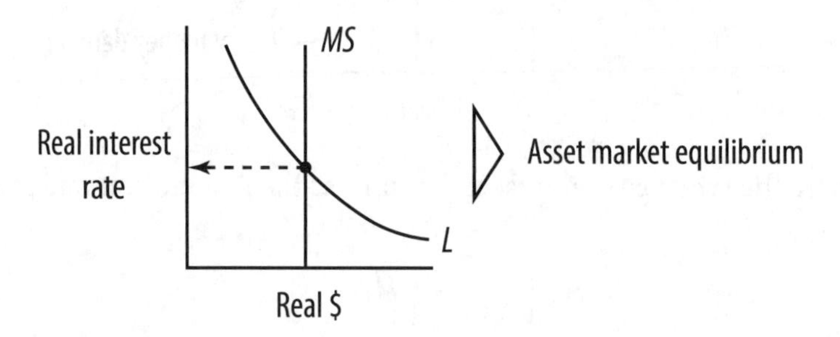


## LM CURVE

Almost there. What happens to the asset market if national income *Y* increases? Here's the simple chain of logic to remember.

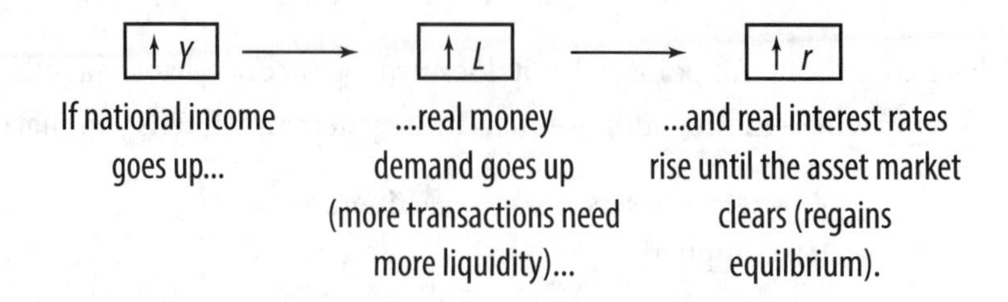

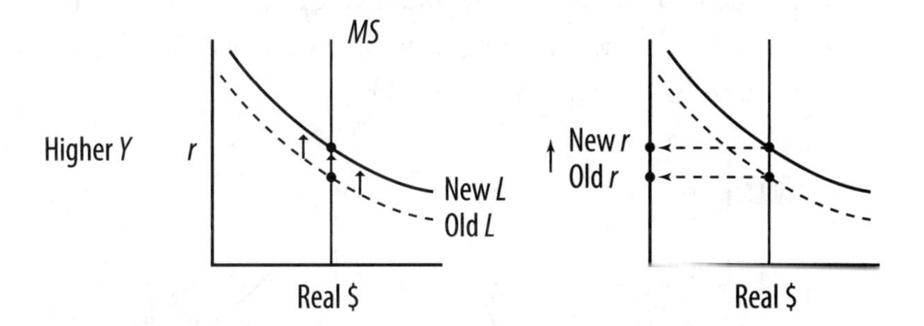


Notice that for any level of national income $Y$, there'll be a certain real interest rate $r$ that clears the asset market—and the higher $Y$ is, the higher the "asset-market clearing" $r$ is. This relationship is called the ***LM curve***. It goes up to the right.

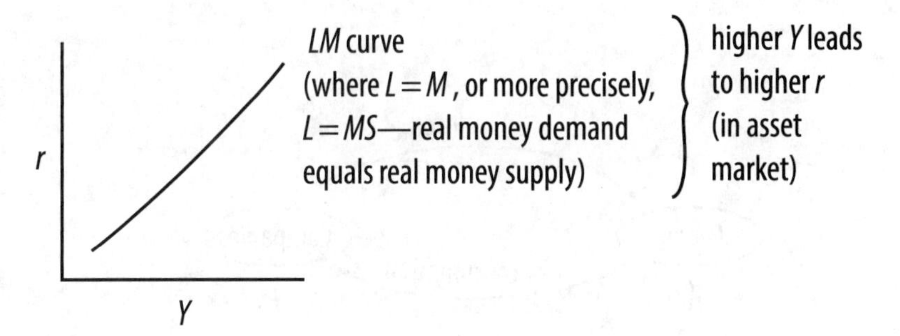


## INVESTMENT AND SAVINGS

The real interest rate $r$ is the master dial on the economic engine. It determines what happens throughout the economy, not just in the market for assets.

Take the market for investable dollars. Since $r$ is the interest paid in general on investments, $r$ is the "price of money" yet again—this time, the price of money for investments (not the literal price of holding cash, as it was earlier).

If you're a company and you can borrow money cheaply (at a low $r$), you'll do so and make more investments in your business. On the other hand, if you can't borrow money cheaply—if the cost of borrowing is high—then you'll borrow less and invest less. So the demand for investment $I^d$ goes down when $r$ goes up, and vice versa. This is a classic demand curve, sloping down to the right.

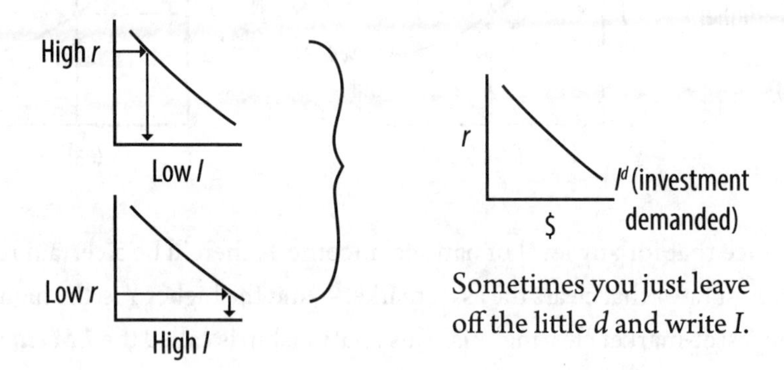


Sometimes you just leave off the little $d$ and write $I$.

Where does the supply of investment money come from? From savings.

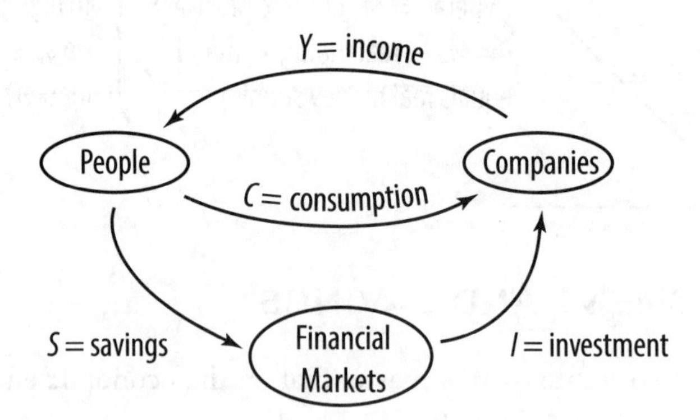


People will save more if they'll get a higher interest rate (a higher price for their saved money). This is a classic supply curve, sloping up to the right.

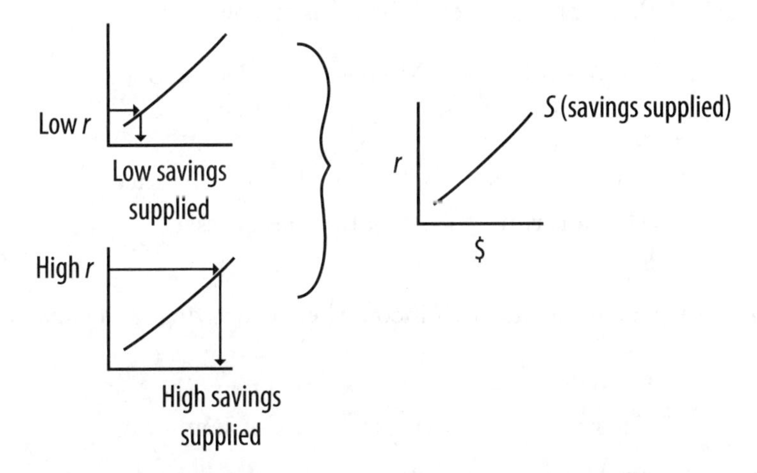


The intersection of these two curves determines the equilibrium in the market for investable dollars:

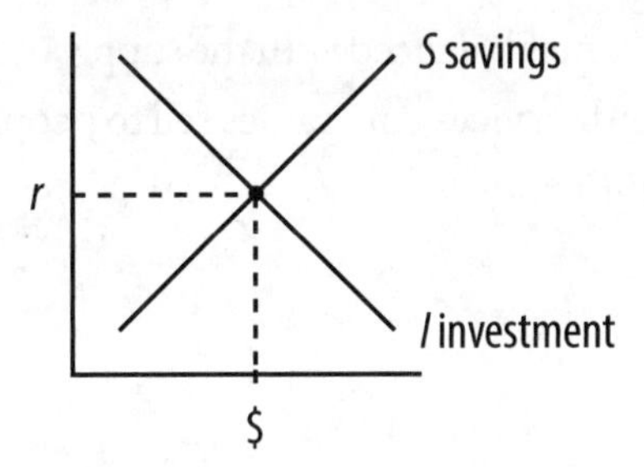


And . . . just as there was a two-fer before, this particular equilibrium gets you a bonus equilibrium as well. Before, equilibrium in the *money* asset market gave you equilibrium in the *non*-money asset market.

| Money demand | = | Money supply | at some interest rate *r* |
|---|---|---|---|
| | ↓ | | |
| Demand for nonmonetary assets | = | Supply of nonmonetary assets | at the same interest rate |

This is because every dollar of wealth is either monetary or nonmonetary, and in the short run, wealth is constant. So if everyone allocates his or her wealth between money and nonmoney so that money equilibrium is reached, then nonmoney equilibrium is reached too.

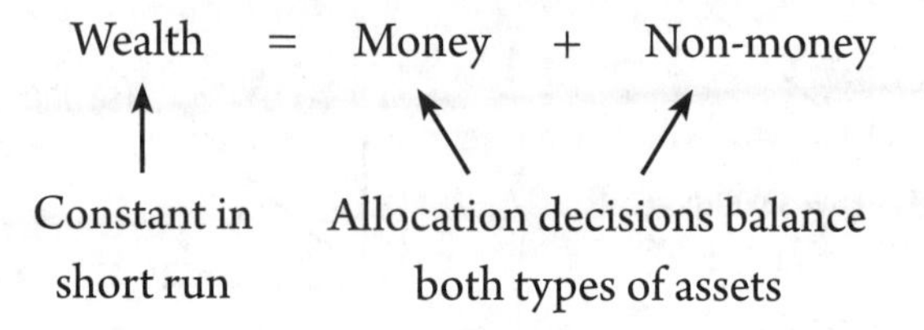


Likewise, people allocate their income between savings and consumption.

Income = Savings + Consumption

$Y$ $\quad$ $S$ $\quad$ $C$

And if savings and investment reach equilibrium at some interest rate *r*, then consumption also reaches equilbrium. That is, the *demand* for consumption C becomes balanced with the supply of stuff to consume—that is, the goods and services companies sell to people. That's called the ***goods market equilibrium***.

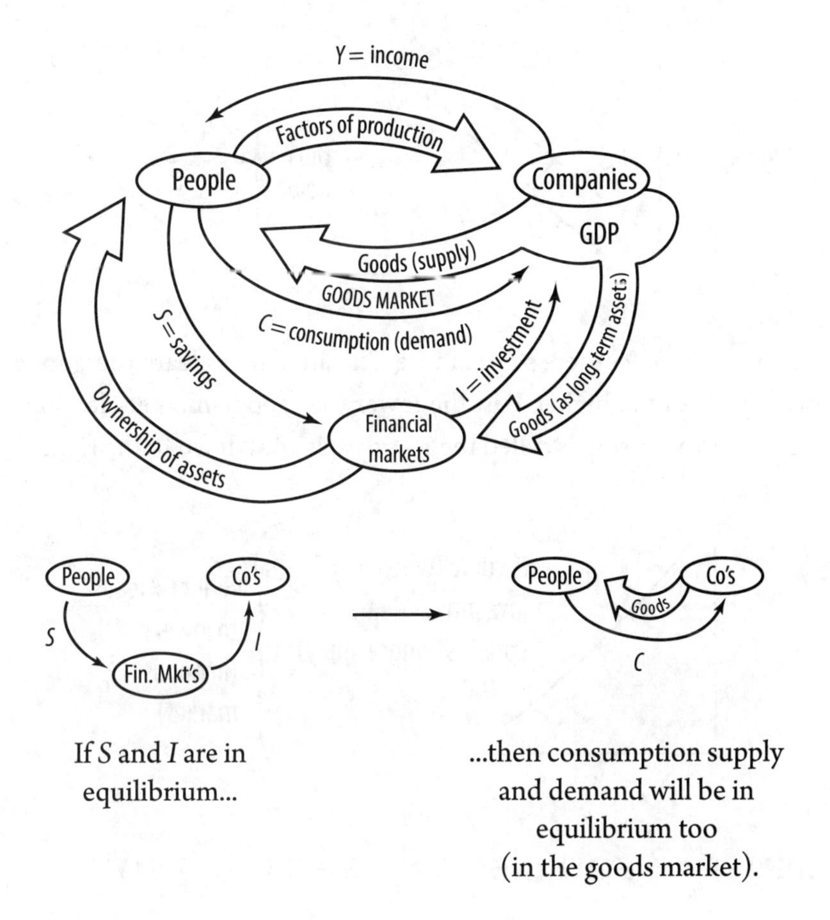


## IS CURVE

What happens to the goods market if national income *Y* increases? As before, here's the simple logic to remember.

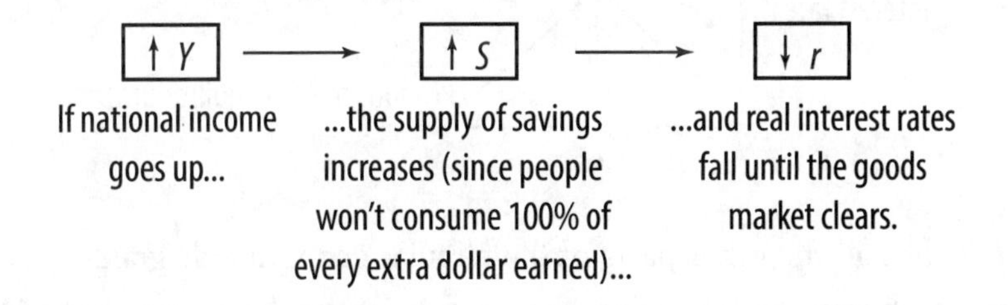

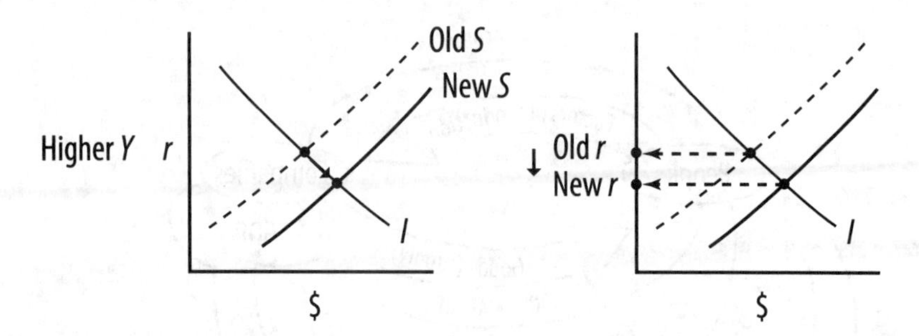


At any level of Y, there's a real interest rate r that clears the goods market—and the higher Y is, the lower the "goods-market-clearing" r is. This relationship is called the ***IS curve***. It goes down to the right.

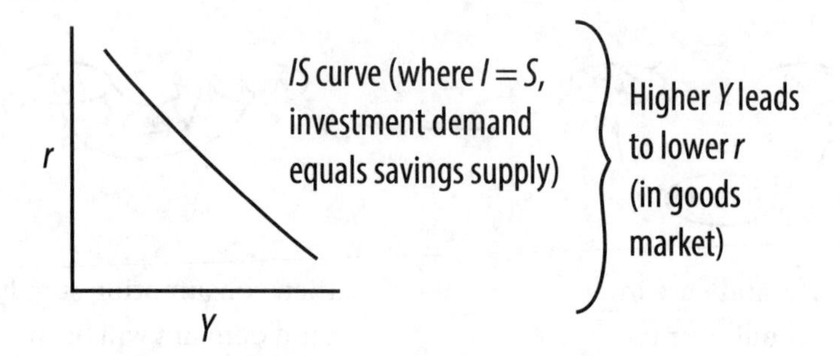


## GENERAL EQUILIBRIUM: IS = LM (= FE!)

The big kahuna picture rolls this all together.

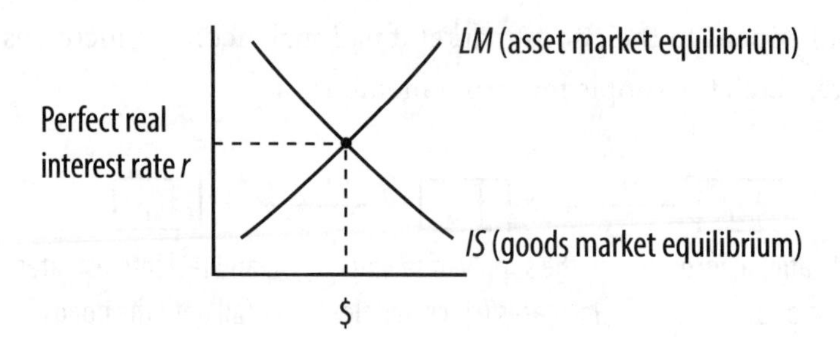


And since all the other parts of the circular economic diagram are in balance, humming along fine, the *top* part of the diagram must be in equilibrium too. This is where people work for companies, exchanging their labor and any other factors of production for wages, etc.

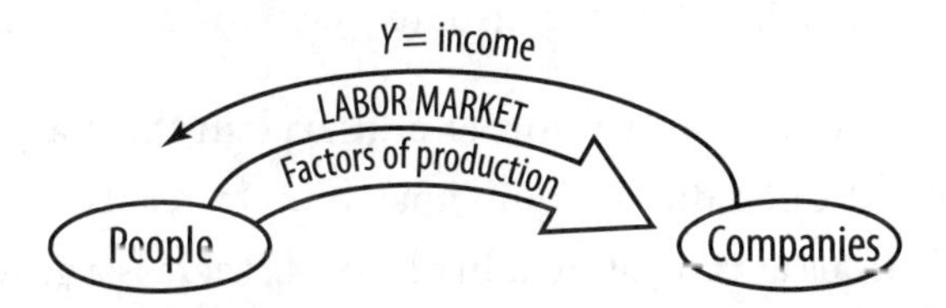


Thus, the economy must be at ***full employment*** (FE)—everyone who wants to work is working productively. So we can throw in one more line—the full employment line, which is fixed for any interest rate.

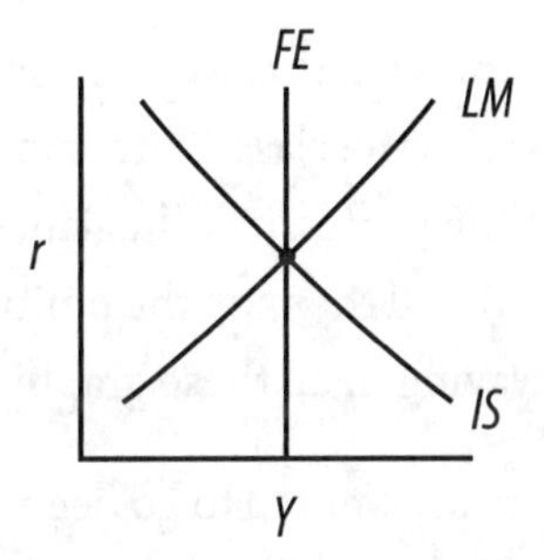


Of course, if achieving full employment were as easy as drawing a vertical line on a graph, the world would be a very different place.

At this stage, you'll start shifting the *IS, LM,* and *FE* curves around in response to various "shocks," such as a sudden influx of workers into the labor market. These shocks will have an effect in the "short run"—one of those ill-defined terms that you can argue about all day long—and also a "long-run" effect ("long-run" being usefully defined either as longer than "short-run" or as the time frame in which we're all dead).

For instance, it can be argued that monetary policy—creating or destroying nominal dollars—only affects these curves in the short run, after which they return to their original equilibrium. But not all economists agree on this. Indeed, infinite are the arguments of macroeconomists, who can at times resemble the high priests of arcane sects rather than scientists. After all, one would think that the great natural experiment of the 2008 financial crisis would, if nothing else, provide reams of additional data and significant insights into the functioning of the global economy. Instead, the academic disputes seem to have only multiplied.

Macroeconomics attempts to order and explain the real world conceptually, but it often doesn't quite match up to reality—particularly where predictions are concerned. In response, there is a growing interest in behavioral economics, which posits that people often behave in idiosyncratic (not always perfectly rational) ways that influence market behavior. However, behavioral economics is not part of traditional b-school economics courses.

In macro, we each sometimes felt that we had to suppress our own thinking if we wanted to do well in the class. Unfortunately, the game to some extent was to figure out the professor's allegiances (e.g., to neoclassical or Keynesian economics) and then toe the political line. There can also be an element of handwaving in all these graphical maneuvers.[6]

Let's be honest—if you really wanted to go deep and hard into this stuff, you'd be getting a master's or a PhD. Figure out what really matters for your grade (e.g., increases in Total Factor Productivity seem to be the cure for almost any macroeconomic ill) and learn to parrot that back, Polly.

## GAME THEORY

The final part of this giant chapter is about ***game theory***. This is what Russell Crowe's *Beautiful Mind* gave birth to.[7] It's the study of specialized interactions called ***games***, which have ***players***. Each player makes a choice or choices, trying to win the most money (or something else good) and/or avoid penalties. This is known as ***maximizing your payoff***.

---

6 We appreciate that macro has staggeringly complex objects of study (i.e., national economies).

7 Okay, it was actually John Nash, played by Russell Crowe, who is actually a Roman gladiator.

The key to all of game theory is to ask yourself one question, punk:

**WWTOPD?**

What Would the Other Player Do?

In other words, assume the other player is as smart and as motivated as you. Look at the game from his or her point of view. Think as far ahead as you can.

## MECHANICS OF GAME THEORY

The first kind of game you'll study (in micro or some other course) is simple:

- ***Two players***: You're player A, and player B is Bob.
- ***Simultaneous***: You both ***move*** (make your choice) at once. Think Rock-Paper-Scissors. You don't have to literally move simultaneously—you could write down your moves on separate pieces of paper. The important thing is that neither of you knows anything about the other player's choice in advance.
- ***Only two choices***: Each of you has a choice between just two possible moves. Rock-Paper-Scissors has three possible moves, so this simple game is even more basic than that.
- ***One shot***: You play once, and then you're done, even if you tie.

The way to represent such an exciting game is with a two-by-two matrix:

| | | Bob (Player B) | |
|---|---|---|---|
| | | Left | Right |
| You (Player A) | Top | | |
| | Bottom | | |

You choose top row or bottom row; Bob chooses left column or right column.

> You: "Are you ready, Bob?"
> Bob: "Ready as I'll ever be."
> You: "Okay, on the count of three. One . . . two . . ."
> You: "Bottom!" Simultaneously: Bob: "Right!"

In this scenario, the ***outcome*** of the game would be the bottom right box.

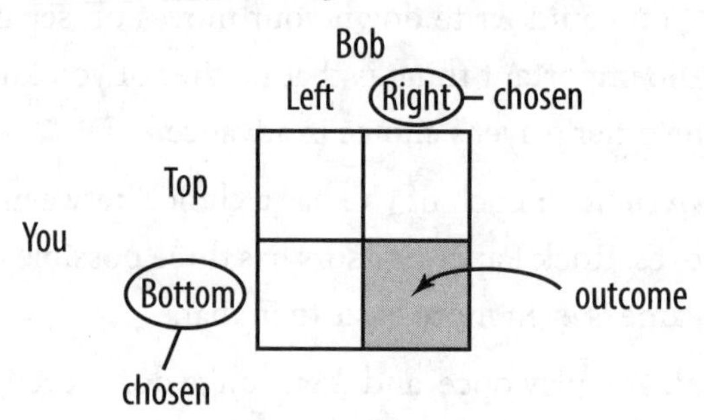


Now you look in that box to see what each player has won from some third party. (Obviously, you were able to see these rewards in advance.)

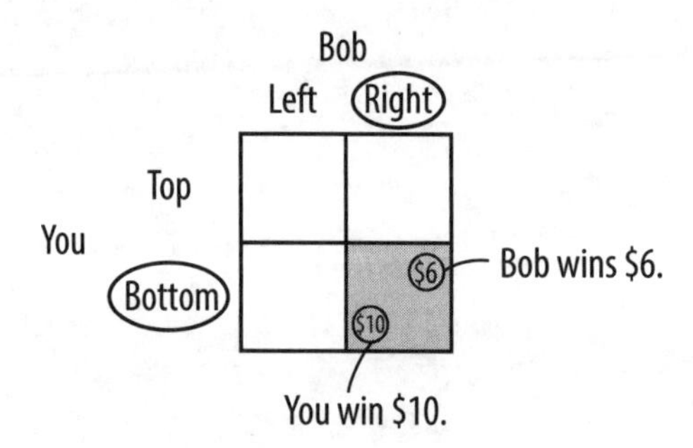

The rewards for you are in the bottom left corner of each little box. Here's a full set of your possible payoffs.

| You | | Left | Right |
|---|---|---|---|
| | Top | $16 | $1 |
| | Bottom | $3 | $10 |

For example, if "Top Left" is the outcome you win $16.

The rewards for Bob are in the top right corner of each little box. Here's a full set of Bob's payoffs:

| | Bob: Left | Bob: Right |
|---|---|---|
| Top | $7 | $12 |
| Bottom | $1 | $6 |

And here's the full game:

| You | | Bob: Left | Bob: Right |
|---|---|---|---|
| | Top | $16, $7 | $1, $12 |
| | Bottom | $3, $1 | $10, $6 |

As you can see, the game is all about the pattern of payoffs. So what would you *really* do in this game? If you just study your own payoffs, you might be puzzled.

| You | | |
|---|---|---|
| Top | $16 | $1 |
| Bottom | $3 | $10 |

You: “Hmm. If I go for top, I might win $16—but then again, I might only get $1. If I go for bottom, I only have an upside of $10, but my downside of $3 is better than $1 ...”

Stop! Remember you’re not playing against a coin toss. You’re playing against *another player*, someone just as smart and motivated as you.

What Would The Other Player Do?

Look at the game through Bob’s eyes. What would Bob do?

| Left | Right |
|---|---|
| $7 | $12 |
| $1 | $6 |

Here’s what Bob thinks, or ought to think:

> “Let’s see. Right looks better at first glance. If A chooses Top row, then I should choose Right because $12 is bigger than $7. If, instead, A chooses Bottom row, then I should again choose Right because $6 is bigger than $1. Either way, Right’s better for me, so I’m choosing Right.”

Notice that Bob only compares his payoffs in the same row.

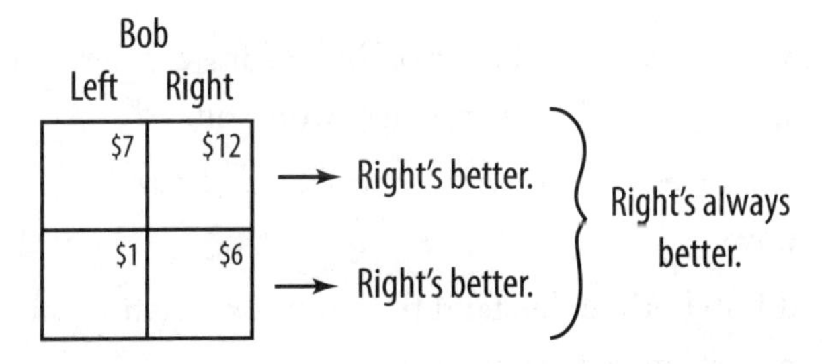


So for Bob, Right is a ***dominant*** strategy, because no matter what you do, Right is better for Bob than Left.[8]

Now come back to being player A. Knowing that Bob is going to choose Right, because that's his dominant strategy, what should *you* do? Look at *your* payoffs.

| You | | Left | (Right) |
|---|---|---|---|
| | Top | $16 | $1 |
| | Bottom | $3 | $10 |

$10 is bigger than $1, so you should choose Bottom. Put away your wishful thinking—you aren't getting that $16.

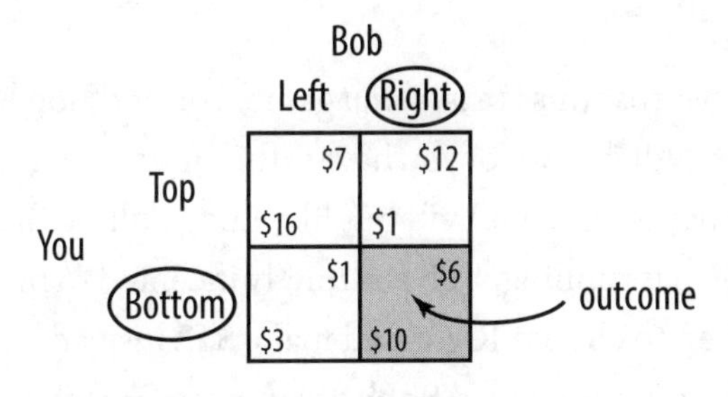


8 This concept of dominance can also be applied in real life situations. Say you're comparing several job offers (lucky you). If some job is worse than another particular job on every dimension, the worse job is *dominated.* A *dominant* job is better than every other on every dimension. Take the dominant job offer, not any dominated one.

This outcome is known as a ***Nash equilibrium***. Here's what defines a Nash equilibrium:

- You can't make yourself better off by making a different move by yourself. If you switch to Top, you're worse off.
- Likewise, Bob can't improve his payoff by *unilaterally* making a different move.
- You and Bob both understand this. Your expectations about each other's actions match reality.

What's interesting about this game is that there's a better outcome on the board—for *both* of you—but there's no way to get there.

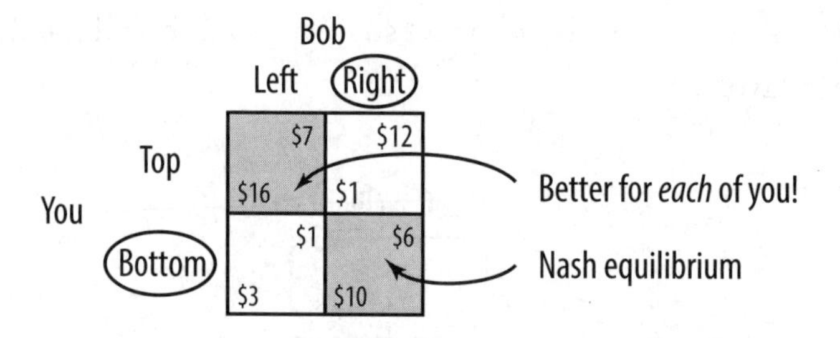


If only you and Bob could agree ahead of time! You want Bob to choose Left, and you'll choose Top. You'll make $16 instead of $10, and Bob will make $7 instead of $6 ... maybe you even offer Bob a kickback ahead of time, since you're going to make a lot more, whereas the improvement's minor for Bob.

The problem is that this is a *one-shot* game. You and Bob have no history and no future. All the two of you have is this one moment. So Bob might promise all day long to go Left, but like Lucy pulling the football once Charlie Brown is running, Bob is strongly incented to make you choose Top (and then to choose Right and make $12 instead of $7). Knowing this, you should not believe Bob's assurances. You should go Bottom. Bob will go Right. And you're back at the Nash equilibrium.

The classic ***Prisoner's Dilemma*** game has slightly different payoffs but the same problem overall: the Nash equilbrium is not the best outcome on the board.

| | | Prisoner B | |
|---|---|---|---|
| | | Cooperate | Defect |
| Prisoner A | Cooperate | \$6<br>\$6 | \$10<br>\$0 |
| | Defect | \$0<br>\$10 | \$3<br>\$3 |

For each prisoner, *Defecting* is the dominant strategy, so the Nash equilibrium is in the Bottom Right again. By the way, the names of the moves can be confusing. *Cooperate* means "cooperate *with each other* and stand strong against the authorities who are trying to get the prisoners to rat each other out." Likewise, *defect* means "defect from the coalition with the other prisoner," in other words, "betray the other prisoner by snitching."[9]

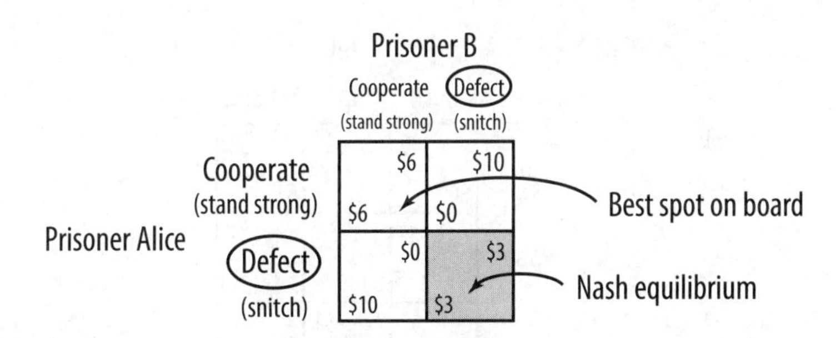


The Prisoner's Dilemma provides a good model of a cartel—and why cartels fail. Producers in a cartel have strong incentives to defect, so the arrangement falls apart unless enforced by other means.

---

9 To make matters more confusing, sometimes the Prisoner's Dilemma is written with *years of prison time* as the payoffs. Since you want *less* prison time, you have to make sure you're going for *low* numbers (not high ones, as you would with dollars).

The Prisoner's Dilemma also provides a counterexample to the ***invisible hand***, Adam Smith's famous image of how the individually self-interested actions of buyers and sellers in a free market lead to the best outcome overall. In the Prisoner's Dilemma, each prisoner wisely acts in his or her own self-interest, but the result is not the best one possible.

A good way out of the Prisoner's Dilemma is to repeat the game for many rounds. In a ***repeated game***, your reputation matters, and you'll cooperate more often (if only in your own self-interest). In real life, few games are truly one-shot. If you screw your partners on one deal, you might not see many other deals—or you might only find very suspicious partners in the future.

## BE UNPREDICTABLE

The best strategy in a repeated game can be a ***mixed strategy***, in which you change up your moves. The most important thing is to be unpredictable, because the other players will exploit any patterns in your moves.

Consider Rock-Paper-Scissors with the following payoff structure:

| | | Bob | | |
|---|---|---|---|---|
| | | Rock | Paper | Scissors |
| You | Rock | $0<br>$0 | $1<br>–$1 | –$1<br>$1 |
| | Paper | –$1<br>$1 | $0<br>$0 | $1<br>–$1 |
| | Scissors | $1<br>–$1 | –$1<br>$1 | $0<br>$0 |

Every time you win, Bob pays you a dollar; when you lose, you pay him a dollar. This is an example of a ***zero-sum game***: your gain is someone else's loss (and vice versa).

So what's your best opening move? If you don't know anything about Bob's past play, then you should roll dice and pick completely at random.

Your chance of picking Rock: ⅓
Your chance of picking Paper: ⅓
Your chance of picking Scissors: ⅓

That's also Bob's best play, if he doesn't know anything about *your* history of playing Rock-Paper-Scissors. You and Bob will be evenly matched, and your expected winnings over time will be $0.

If Bob tends to throw Rock more often, then you can study the percentages and adjust your own probability accordingly (you'll throw Paper more often). This way, your expected winnings over time will be above $0.

By the way, to be truly unpredictable, actually roll dice, flip a coin, or use a random-number generator in Excel. Don't just make up what you believe to be a random string of moves. You're likely to flip-flop too often. Truly random strings contain significant runs of just one move.

## EVEN BE IRRATIONAL (SOMETIMES)

A counterintuitive result of game theory is that it can be smart to be crazy. Take the game of Chicken, in which you drive a car straight at Bob, who's driving another car straight at you. Whoever swerves first loses and is a loser. It's not exactly a one-shot simultaneous game, but let's look at it that way for now:

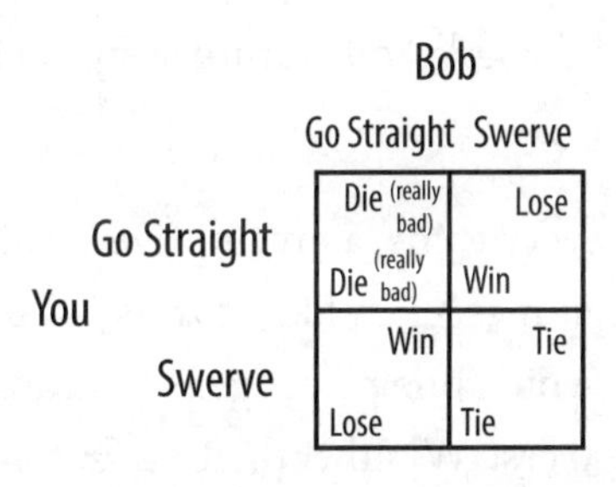


Crazy to even play, right? Say that winning is worth $10 million. If you lose, you fork over half a million dollars. A tie is worth $0. And dying is still really bad. You might play just for the chance at that ten mill.

Here's how to win.

As quickly as you can, snap off your steering wheel, wave it out the window so Bob can see it, then throw it away.

You have now ***strategically committed*** to your course of action. You have removed your own fallback, play-it-safe option. Now here's what Bob faces:

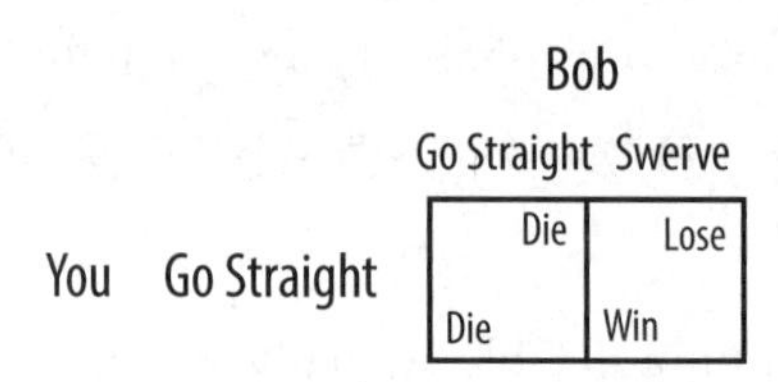


Bob now can only choose to lose or to die. Being rational, he will choose to lose, and you will win. You just have to make sure you break your steering wheel *before* Bob does—*not at the same time.*

This concept of ***strategic commitment***—tying your own hands, cutting off your own escape routes, burning your own ships[10]—may seem utterly nuts. After all, isn't having choices good? Not always. By restricting your own choices, you can force Bob's hand. This even works when Bob is you as well—your future self, say. If you want to make sure you get up early tomorrow, set your alarm, but then take away the snooze button option. Place the alarm clock across the room. Or ask a sadistic roommate or two to wake you.[11] Your future sleepy self won't be able to betray your intentions.

This is not to say that in every game, a preemptive, bold move is always the right thing to do. Sometimes "crazy like a fox" is just plain crazy. Some games are ***sequential***, in which one player goes first, and it's not always the case that you should go first. Whether there's a ***first-mover advantage***

10 As Hernán Cortés supposedly did after he landed on the shores of Mexico.

11 One of us did this once in undergrad. The roommates used ice, shaving cream, and a compressed air horn, to impressive effect.

or a ***second-mover advantage*** depends on the game. The point is that you should really understand the game—its structure, its payoffs—in order to figure out your best play, which may or may not involve:

- *Seemingly irrational behavior*, such as cutting off your own options.
- *Random, unpredictable behavior*, such as choosing your next move by a coin flip.

## GOING ONCE, GOING TWICE...

***Auctions*** are also an object of game-theoretic study. You probably already know about ***English auctions***, in which bidders make higher and higher offers until there's only one bidder left. This is the kind of auction you run with a fast-talkin' auctioneer:

"... I've got $500, $500 from the little lady in the first row, do I hear $600, $600, going once, $600 from the gentleman in the cowboy hat, $600 for this fine oil painting of Elvis Presley's Bluetick hound, $600 going once ..."

So you bid $700 for this painting and *wham*! you won. You were the highest bidder, and now you pay $700. As you gaze at the sad-eyed Bluetick, you wonder, "No one else bid as high as me. Maybe it's not actually worth $700. Maybe I could have gotten it for $650 ..." This is known as the ***winner's curse***. In an English auction, if you win, you tend to suspect you overpaid. There's also a lot of gamesmanship in an English auction—as a bidder, your best strategy depends not just on your own private valuation of the painting, but also on how many other bidders there are, how they behave, etc.

In contrast, a ***second-price auction*** (also known as a ***Vickrey auction***) removes all the gamesmanship at a stroke. Every bidder submits one sealed bid. The highest bid wins—but that bidder only pays the *second* highest bid.

Amazingly, because of this change in the rules, your best strategy as a bidder is to submit your true valuation—your indifference point—or something just below it, so that you're *just* still willing to buy. You no longer care what the other bidders do.

Why does the second-price auction work so well? Let's say you really think that the painting is worth $675, or rather you'd just be willing to pay $675. You bid $675, and the next highest bid is $650. You win, and you pay $650. Should you have bid any lower than $675, say $655? No, because you'd still only pay $650. And it increases the chance that some *other* bidder could have bid lower than $675 *and won,* when you would have bought at $675. In a second-price auction, lowering your own bid to try to snag a bargain doesn't work, because it doesn't reduce the price you pay if you win. All it does is increase the chance that you lose the auction to a bidder who bid *less* than you were willing to pay. A similar argument explains why you wouldn't overbid either.

Second-price auctions, with modifications, have been adopted in various real-world scenarios, including the sale of Treasury bills. The initial public offering of Google was a kind of second-price auction.

## I CALL THAT A BARGAIN

You might take a whole elective course on bargaining and negotiations, but in case you encounter it within a first-year course, remember three things:

1. As in game theory, look at the situation through the other side's eyes. Assume they're as smart as you.
2. Figure out your own ***BATNA***—the ***best alternative to negotiated agreement***. In other words, what's your best option if you can't reach agreement? By improving your own BATNA, you strengthen your hand in the negotiation.

1. + 2. Think about the other side's BATNA as well.

3. Never forget the ***larger game***—that of business school. If you win a classroom negotiation in a dirty way and become known as a giant jackass, you have lost points in that larger game. Of course, play to win, but always be a good sport.

# Chapter 3: Statistics

Statistics is a rough but necessary subject. If you have a lot of quantitative information, stats help you boil it down to a few key numbers so you can make predictions and decisions. The word *statistics* can refer either to the subject ("Stats is fun!") or the key numbers themselves ("These performance statistics suck").

There are five big areas of stats to cover:

| | | |
|---|---|---|
| 1. | Descriptive Statistics | "The average salary is \$X..." |
| 2. | Probability | "The chance of failure is 5%..." |
| 3. | Distributions | Bell curve |
| 4. | Sampling & Hypothesis Testing | "Given the sample of 100 customers we can conclude..." |
| 5. | Correlation & Regression | "For every extra million dollars we spend on advertising, our market share increases by 3%..." |

## DESCRIPTIVE STATISTICS

Say there are 500 people in your business-school class, and you want to think about the *number of years* each of you spent working between college graduation and business school.

| Student | Years Since College |
| --- | --- |
| You | 4 |
| Alice Atwater | 2 |
| Bill Burns | 6 |
| ... (497 more) | ... (497 more) |

Each row is an ***observation.***

To make things simple, you'll probably round to the nearest whole number (instead of having data like 5.25 years, 7.8 years, etc.). Whole numbers are ***discrete*** (meaning "separated and countable"), so with this information, you can make a ***histogram*** to display the count in each category.

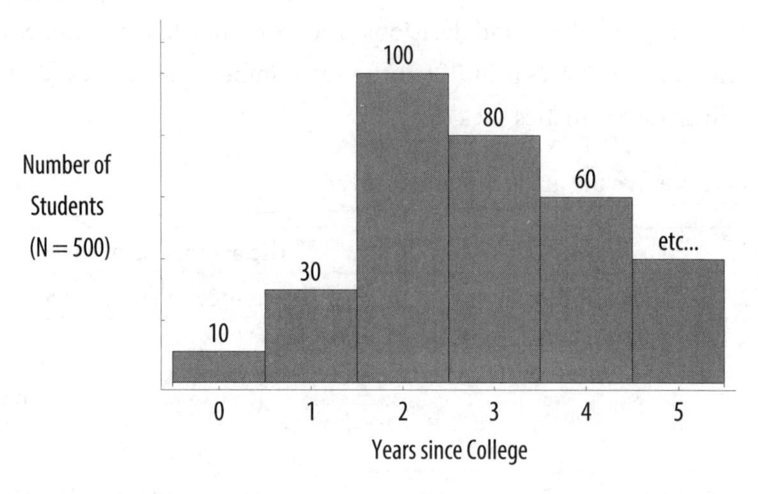

If you convert to percents, you can also show the same graph as a ***frequency distribution***.

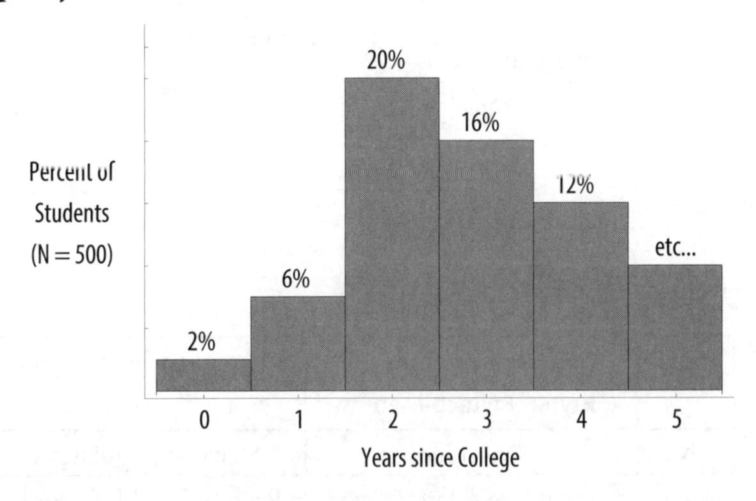


This pretty picture illustrates a link between statistics and ***probability***. If you pick someone at random from your class, there's a 2% chance that he or she spent zero years working, a 6% chance he or she spent one year, etc.

Now, what's the ***average*** amount of time your classmates have spent in the real world since college? There are three primary ways to answer this question:

1. Mean
2. Median
3. Mode

} Three types of average

The ***mean*** is most important. In fact, this is what Excel means by the (AVERAGE) () function. Technically, this is the "arithmetic mean" (air-ith-MET-ik), but you'll never use the other means defined by statisticians, so you can just say "mean." You already know the formula from the GMAT or GRE:

$$\text{Mean} = \frac{\text{Sum}}{\text{Number}}$$

You add up everyone's "years since college" and divide that total by 500, the number of people in your class.

$$\text{Mean Years} = \frac{\text{Total Years}}{\text{Number of Students}}$$

$$3.36 = \frac{1{,}680}{500}$$

Here are some key symbols.

| | | |
|---|---|---|
| $\mu$ | Greek letter *mu* ("myoo") which is an "m" | Mean of a whole population (like your class) |
| $N$ | Number | Size of a population |
| $\Sigma$ | Capital Greek letter *sigma,* which is a capital "S" | Process of adding numbers to get a sum |

Since $\Sigma$ represents a process, you have to tell it what to do. First, to label the different values you want to add up, use *subscripts* (like the 1 in $x_1$).

| **Student** | **Years Since College** | |
|---|---|---|
| You | 4 | $= x_1$ |
| Alice Atwater | 2 | $= x_2$ |
| Bill Burns | 6 | $= x_3$ |
| ... | ... | ... |

So now you can write the Total Years this way:

$$\text{Total Years} = x_1 + x_2 + x_3 + \ldots + x_{500}$$

But that's really annoying. Instead, use $\Sigma$ to indicate that you're summing all the $x$'s up from $x_1$ through $x_{500}$.

$$x_1 + x_2 + x_3 + \ldots + x_{500} = \sum_{i=1}^{500} x_i \longleftarrow \text{dummy subscript}$$

A curly *i* or *j* is a typical "dummy" subscript in the expression.

$\sum_{i=1}^{500} x_i$ means "add up all the $x$'s, starting at $x_1$ and going up to and including $x_{500}$." The $i = 1$ on the bottom means "start with $x_1$." The 500 on top means "finish with $x_{500}$."

Once it's all Greeked up and generalized, the mean equation looks like:

$$\mu = \frac{\sum_{i=1}^{N} x_i}{N}$$

But even in a toga, this equation is still just:

$$\text{Mean} = \frac{\text{Sum}}{\text{Number}}$$

You can compute the mean "years since college" another way. Since a lot of the numbers are repeated, it makes sense to add them up in groups.

Add ten 0's
thirty 1's,
a hundred 2's,
eighty 3's,
sixty 4's, etc…

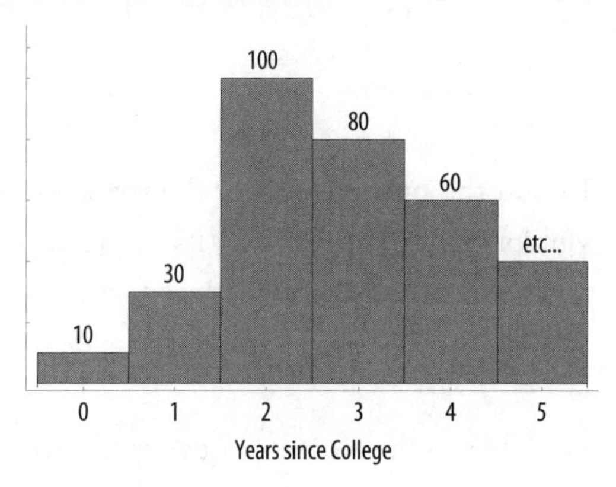

Ten 0's is ten times zero. Rewriting, you get:

$$\text{Mean} = \frac{10 \cdot 0 + 30 \cdot 1 + 100 \cdot 2 + 80 \cdot 3 + \ldots}{500}$$

Now split up the numerator:

$$\text{Mean} = \frac{10}{500} \cdot 0 + \frac{30}{500} \cdot 1 + \frac{100}{500} \cdot 2 + \frac{80}{500} \cdot 3 + \ldots$$

$$= 2\% \cdot 0 + 6\% \cdot 1 + 20\% \cdot 2 + 16\% \cdot 3$$

Notice that the percents you just calculated are the same as those on the frequency distribution.

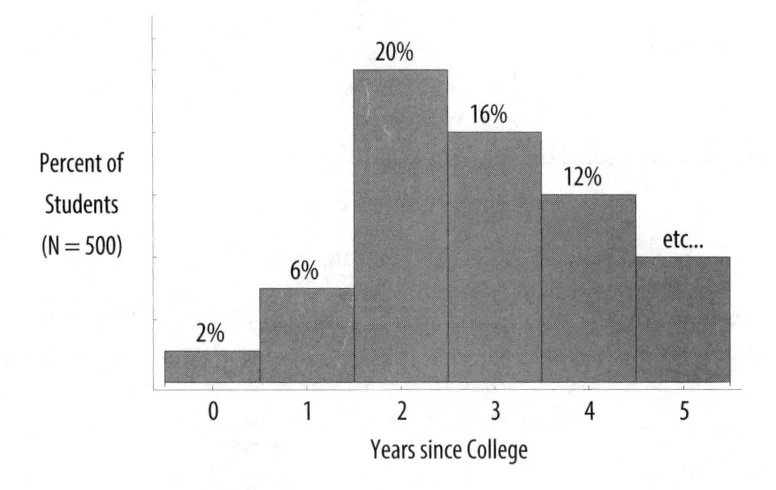


To find the mean, consider the ***frequencies*** as percents or decimals. Multiply each observation by its frequency, and add the results up. You'll get the mean. In toga form, the equation looks like:

$$\text{Mean} = \Sigma x \cdot p(x)$$

where $p(x)$ is the percent frequency of $x$ occurring in the population—in other words, $p(x)$ is the *probability* of picking someone at random with $x$ years since college. Include every possible value of $x$ in your sum.

Written this way, the mean is also called the ***expected value*** of $x$. It's the "average" value you'd expect if you pulled a lot of people at random and averaged their $x$'s (years since college). Expected value is not the value you'd expect from any one person necessarily. It's like when the census reports that the average US family with children has 1.9 children. We've never met 0.9 of a child, but 1.9 is still the expected value considering the country as a whole.

The ***median*** is the middle number, or the 50th percentile: half the people have more years since college (or the same number), and half have fewer years (or the same number). You can read the median from the percent histogram—just add from the left until you hit 50%.

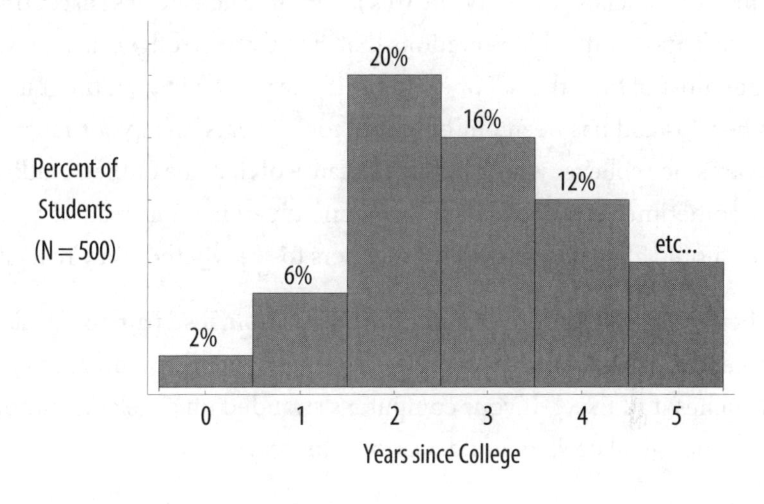


| Years since College (at least) | | Cumulative Percent |
|---|---|---|
| 0 | | 2% |
| 1 | + 6% = | 8% |
| 2 | + 20% = | 28% |
| 3 | + 16% = | 44% |
| 4 | + 12% = | 56% STOP! |

This means that the median is four years. The ***mode*** is the observation that shows up most often, corresponding to the highest frequency on the histogram. If none of the years to the right of 4 have more than 20% of the population, then the histogram's peak is 20%, and the mode is two years.

## GIMME THE SPREAD

Mean, median, and mode are all "central" measures—they answer the question "where's the center of all the data?" But often, you want to know how spread out the data is.

The crudest measure of spread is ***range***, which is just the largest value minus the smallest value. While this is easy to calculate, it's susceptible to ***outliers***—oddball observations that, rightly or wrongly, lie far away from most of the others. For example, if one person in your program is in her 70s and has been out of school for 50 years, then your range of "years since college" would be huge because of that one outlier. Outliers are sometimes erroneous data (someone typed in 55 instead of 5), so you should always look closely at outliers to see whether they're legit.

A better measure of spread is ***standard deviation***. It's a pain to calculate by hand, but it's got nice mathematical properties, and now everyone just crunches it in Excel. If your computer's stranded on an island without you, you calculate standard deviations this way:

1. Figure out the mean.

| Student | Years Since College |
|---|---|
| You | 4 |
| Alice Atwater | 2 |
| Bill Burns | 6 |
| ... | ... |

Mean $\mu = 3.36$

2. Subtract the mean from each observation and square those differences, also known as **deviations**.

| **Student** | **Years since College** | **Minus the Mean** | **Deviation** | **Squared Deviation** |
|---|---|---|---|---|
| You | 4 | $-3.36$ | $=0.64$ | $(0.64)^2=0.4096$ |
| Alice A. | 2 | $-3.36$ | $=-1.36$ | $(-1.36)^2=1.8496$ |
| Bill B. | 6 | $-3.36$ | $=2.64$ | $(2.64)^2=6.9696$ |
| ... | ... | ... | ... | ... |

3. Take the mean of all those squared deviations. That is, add them all up and divide by the number of observations, in this case 500.

$$\frac{0.4096+1.8496+6.9696+\ldots}{500}=7.672$$

This result is known as the ***variance***. Why? It just is, son.

4. Finally, take the square root of the variance.

$$\sqrt{7.672}\approx 2.77=\text{standard deviation (SD)}$$

What the hell does that result of 2.77 mean? Roughly, SD indicates how far on average a data point is from the mean, whether above or below (which would be average distance). That's not the precise mathematical definition of SD, but it's close enough—and average distance is much easier to grasp.

To get average distance, you would measure each point's distance from the mean, ignoring direction. Then you'd take the mean of all those distances (add 'em up and divide by the number of points). Again, although the average distance and standard deviation are not exactly the same thing, they behave in much the same way mathematically, and they're close enough for government work.

Consider a few cases with the same mean, but different spreads.

Case 1: Every observation = 4

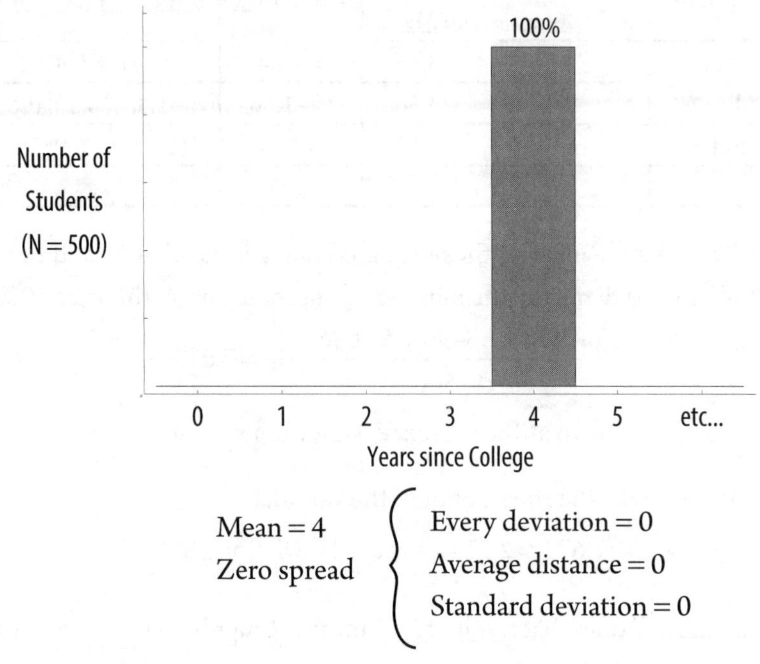


Case 2: Half of the observations = 5, the other half = 3

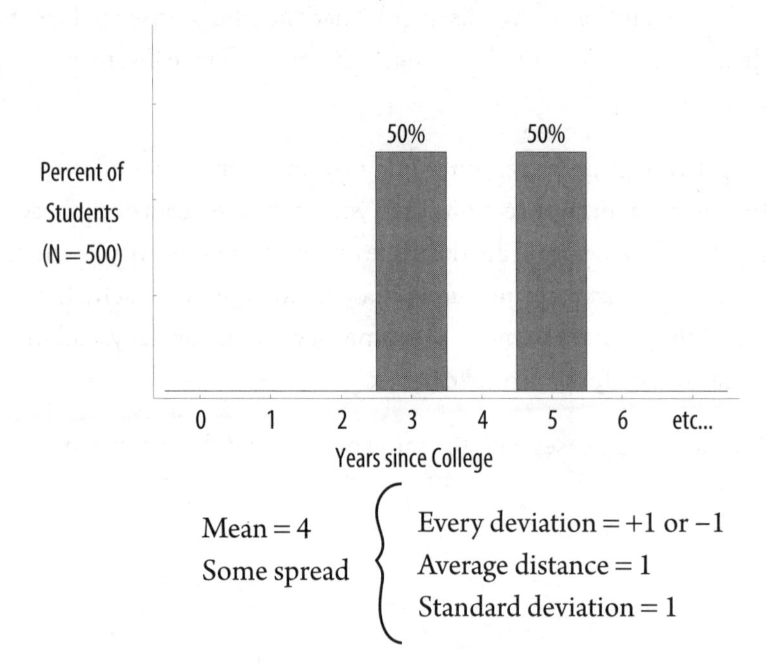

Case 3: Realistic spread

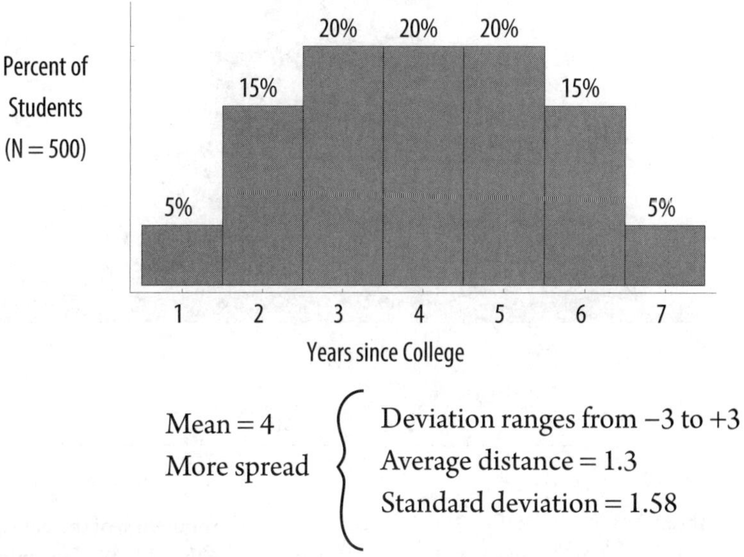


Here, the average distance of every point from 4 and the standard deviation (defined by the weird procedure earlier) are not exactly the same, but they are pretty close. Notice that in the last case, the numbers are not "4 plus or minus 1.58." Usually there's a significant amount of data *more* than a standard deviation away from the mean. But it can be shown that at least 3/4 of the data is always within two standard deviations of the mean, and when the histogram is ***bell-shaped*** (with one central *hump* and two little *tails* like a bell seen from the side), you can make even tighter claims:

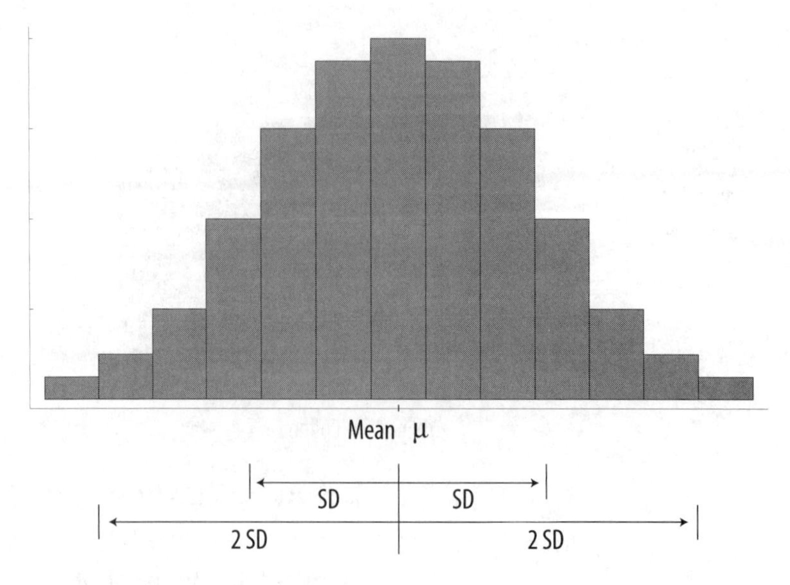


About 2/3 of the data is within 1 SD of the mean

About 95% of the data is within 2 SD's of the mean

If the mean is 4 and the SD is 1.2 for some bell-shaped data, then at least 19 of every 20 observations (95%) are between 1.6 $(=4-2\times 1.2)$ and 6.4 $(=4+2\times 1.2)$.

Standard deviation is incredibly important in other subjects you'll encounter in business school. In finance, standard deviation is a typical way to measure the ***risk*** or ***volatility*** of an investment. You measure the daily percent changes (***returns***) of Apple stock, say, then you compute the standard deviation of a whole bunch of those daily returns. The higher the standard deviation, the more volatile the stock is and the riskier it's considered, generally speaking.

In operations and manufacturing, standard deviation is used in ***process control***. For instance, if you're making lug nuts for nuclear missiles, you'll have tight specifications to meet. The more the lug nuts vary in size, the less likely it is that they'll be acceptable to the government (your client). So you'll have to measure and try to reduce the standard deviation of the width, the thickness, etc., of the batches of lug nuts.

The symbol for standard deviation is $\sigma$, a lowercase Greek letter *sigma*. Don't confuse this with big sigma ($\Sigma$), which means summation. By the way, $\sigma$ is the sigma in the term ***Six Sigma***, a famous process improvement methodology.[1]

Now, what's been assumed so far is that you've been working with the entire population you care about—the 500 people in your b-school class. Often, though, you can't get data from everyone. Maybe you can only get a sample of 50 people, but you'll want to extrapolate your results to all 500.

This difference between ***samples*** and ***populations*** turns out to be so important that the symbols for the statistics themselves are different in each case.

| | Population | Sample |
|---|---|---|
| Size | $N$ | $n$ |
| Mean | $\mu$ | $\overline{x}$ |
| Standard Deviation | $\sigma$ | $s$ |
| Variance | $\sigma^2$ | $s^2$ |

The bar over the $x$ means the average of the $x$'s.

1 Experts in this methodology are called Six Sigma Black Belts, the goofiest martial artists known to man. If you encounter such a nerdy ninja in a dark alley, fear not.

For a sample, the statistics aren't written with Greek letters. You compute the ***sample mean*** $\overline{x}$ ("the mean of a sample set of data") the same way as you do a population mean $\mu$:

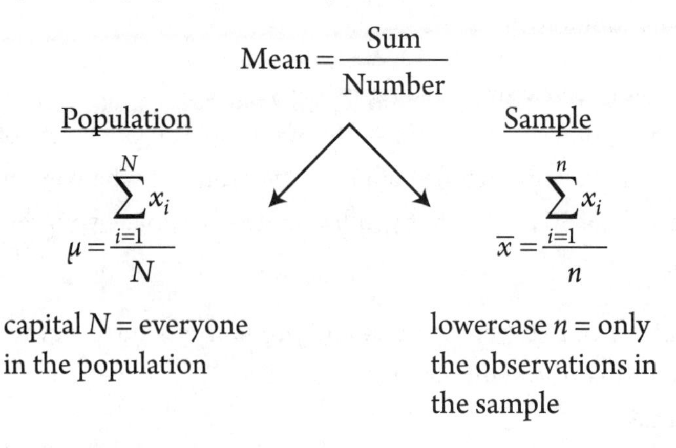


Weirdly, you compute variance and standard deviation slightly differently in each case:

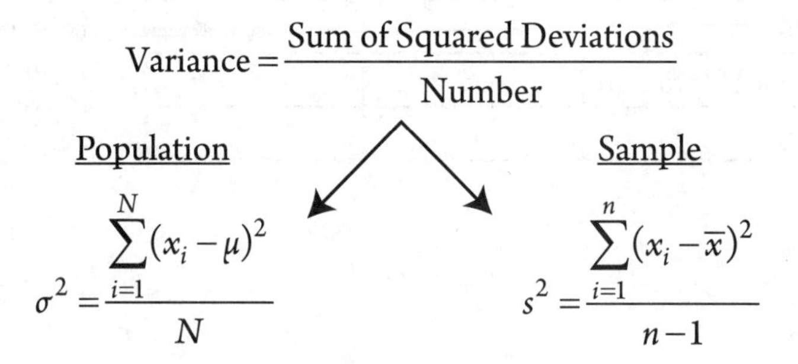


You divide by $n - 1$ on the right, rather than by $n$ (the number of observations in the sample) as you'd expect. The reason is technical, but the rationale is simple: you always use *sample statistics* to estimate *population statistics.* You never care about the sample itself—you compute the sample stats so you can learn about the whole population. And it turns out that, mathematically, dividing by $n - 1$ makes $s^2$ a better ***estimator*** of $\sigma^2$ than dividing by $n$. Why? It just does.

You turn variance into standard deviation the same way for both sides—by taking the square root.

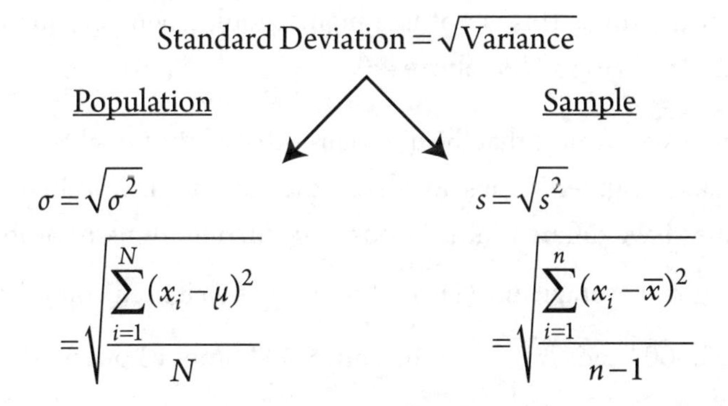


Mean and standard deviation are the two most important statistics to compute about any data set. There are ways to measure ***skew*** (lack of symmetry around the mean) and other aspects of the data, but don't worry about those for now.

## PROBABILITY

You already know that:

$$\text{Probability of Success} = \frac{\text{Number of Successful Possible Outcomes}}{\text{Total Number of Possible Outcomes}}$$

You encountered this on the GMAT or GRE: "What's the probability of rolling an odd number with a fair six-sided die?"

Answer: You could roll any of three odd numbers $(1, 3, \text{or } 5)$, and there are six total possible outcomes $(1, 2, 3, 4, 5, \text{or } 6)$. So the probability is $\frac{3}{6} = \frac{1}{2}$. In math symbols, you write $P(\text{Odd}) = \frac{1}{2}$ or $P(X = \text{odd}) = \frac{1}{2}$.

The $P$ can be upper or lower case. $X$ is a ***random variable***—in this case, the roll of the die.

Probabilities range from 0 (impossible) to 1 (completely certain). You can't have less than 0 success, and you can't be more successful than 100% of the time. This is not like high school, when you might have been able to swing a GPA above 4.0.

Of course, you assume that the die is fair—that all the possible outcomes are equally weighted. That's not always the case in reality. When various outcomes have different likelihoods of occurring, think of probability as a long-run average. If $P(\text{Heads}) = \frac{3}{5}$ for a weighted coin, then you expect 3,000 heads if you flip the coin 5,000 times (3 out of 5).

Probabilities are often written as decimals or percents.

$$P(\text{Heads}) = \frac{3}{5} = 0.60 = 60\%$$

Decimals are the least intuitive. Get used to them; they show up. Also be ready and able to switch to percents or fractions, rephrasing as a long-run average rate.

$$P(\text{Heads}) = 0.60 \longrightarrow P(\text{Heads}) = 60\% = \frac{60}{100} = \frac{3}{5}$$

"60% of the time, I get heads. 3 flips out of 5, I get heads."

By making the probabilities concrete in this way, you'll make more sense of them.

## A MEDICAL EXAMPLE

The complexities of probability are often illustrated through some story about a dread disease that's very hard to detect in its early stages. As the patient, you either have this disease or you don't; there is no middle ground. In techier terms, your true disease state is Sick or Okay.

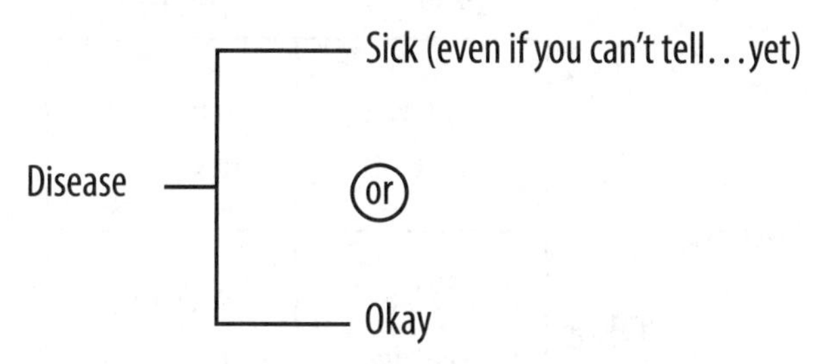


Now, along comes a new test for this disease. If you take this test, you get either a ***positive*** result (supposedly indicating that you have the disease) or a ***negative*** result (supposedly indicating that you don't have the disease).

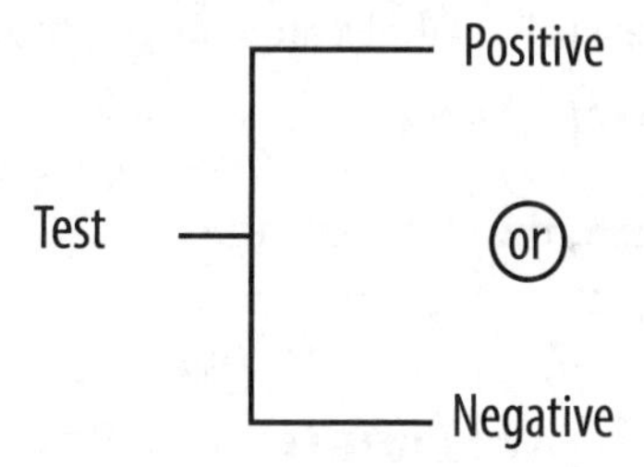


Note that positive in this context is bad, while negative is good.

No test is perfect. There are two kinds of ***false*** (= untrue) results.

- A ***false positive*** is a positive result for someone who *doesn't* actually have the disease.
- A ***false negative*** is a negative result for someone who actually *does* have the disease.

False positives are scary, but false negatives kill people, so if you have to choose just one to reduce, you'd want to reduce false negatives.

You can see all the results in a 2×2 table:

| | | ***Test*** | |
|---|---|---|---|
| | | **Positive** | **Negative** |
| ***Disease*** | **Sick** | True positives | False negatives |
| | **Okay** | False positives | True negatives |

Think through each of these four boxes. One way people mess all this stuff up is by not clearly distinguishing the disease itself from the *test* results. Never use the same labels for the actual disease and for the test. Stick with more descriptive labels, such as the four above (Sick/Okay and Positive/Negative).

You can also total each row and column.

| | | ***Test*** | | |
|---|---|---|---|---|
| | | **Positive** | **Negative** | **Total** |
| ***Disease*** | **Sick** | True positives | False negatives | All sick |
| | **Okay** | False positives | True negatives | All okay |
| | **Total** | All positive | All negative | Everyone |

Something called the ***false positive rate*** has a very counterintuitive definition. People screw this up all the time,[2] so be careful. The false positive rate is the *number of false positives divided by the total number of Okays,* not the total number of positives. So if the false positive rate is 3% on this test, the table will look like this:

| | | ***Test*** | | |
|---|---|---|---|---|
| | | **Positive** | **Negative** | **Total** |
| ***Disease*** | **Sick** | | | |
| | **Okay** | 3 (false pos) | 97 (true neg) | 100 (all okay) |
| | **Total** | | | |

Out of 100 people truly without the disease, 97 people get *true negative* test results, and 3 people get *false positive* results.

$$\text{False Positive Rate} = \frac{\text{False Positives}}{\text{All Okays}} = \frac{\text{False Positives}}{\text{False Positives} + \text{True Negatives}}$$

It's easy to think that true positives go in the denominator, but that's wrong—true *negatives* go there, because they combine with false positives to give you the total disease-free population.

2 In 1999, the *New England Journal of Medicine* published a letter revealing that, out of 63 published stories mentioning "false positive rate" in a two-year period, *nearly half* had miscalculated the number. Maybe you should give this chapter to your friends in med school, or about half of them.

Likewise, a ***false negative rate*** of 2% means that 2% of all sick people get false negative results, while 98% get true positive results.

| | | *Test* | | |
|---|---|---|---|---|
| | | **Positive** | **Negative** | **Total** |
| ***Disease*** | **Sick** | 98 (true pos) | 2 (false neg) | 100 (all sick) |
| | **Okay** | | | |
| | **Total** | | | |

$$\text{False Negative Rate} = \frac{\text{False Negatives}}{\text{All Sick}} = \frac{\text{False Negatives}}{\text{False Negatives} + \text{True Positives}}$$

Let's say that 1% of all people have the disease. This is called the ***base rate***—the natural rate of occurrence of the disease.

Now you have the test done, and the lab results come back:

*Dreadful Disease:* ***Positive***
*Have a nice day!*

*– Dr. Nick*

What's the chance that you *actually* have the disease?

To review, you know three things:

False positive rate = 3%

False negative rate = 2%

Base rate = 1%

To answer the question, you can fill in the grid. The key is to pick smart numbers—large whole numbers that allow you to avoid decimals. For the big total population, you might think that 100 is good. In fact, 10,000 is way better (you'll avoid decimals this way).

| | | *Test* | | |
|---|---|---|---|---|
| | | **Positive** | **Negative** | **Total** |
| *Disease* | **Sick** | | | |
| | **Okay** | | | |
| | **Total** | | | 10,000 |

Now use the base rate. 1% of all people actually have the disease. 1% of 10,000 = 100. That means 9,900 people are Okay.

| | | *Test* | | |
|---|---|---|---|---|
| | | **Positive** | **Negative** | **Total** |
| *Disease* | **Sick** | | | 100 |
| | **Okay** | | | 9,900 |
| | **Total** | | | 10,000 |

Next, use the false positive rate. 3% of Okay people get false positives; the other 97% of Okay people get the negatives. 3% of 9,900 is 297, while 97% of 9,900 is 9,603.

| | **Positive** | **Negative** | **Total** |
|---|---|---|---|
| **Sick** | | | 100 |
| **Okay** | 297 | 9,603 | 9,900 |
| **Total** | | | 10,000 |

The false negative rate states that 2% of Sick people get false negatives, while the other 98% get true positives.

| | **Positive** | **Negative** | **Total** |
|---|---|---|---|
| **Sick** | 98 | 2 | 100 |
| **Okay** | 297 | 9,603 | 9,900 |
| **Total** | | | 10,000 |

Finally, you can sum up the number of positive tests you'll get out of 10,000 people and the number of negative tests.

| | **Positive** | **Negative** | **Total** |
|---|---|---|---|
| **Sick** | 98 | 2 | 100 |
| **Okay** | 297 | 9,603 | 9,900 |
| **Total** | 395 | 9,605 | 10,000 |

Out of the 395 positive test results, only 98 people are *actually* Sick. This means that even if you get a positive test result, you only have a $\frac{98}{395} \approx$ 25% chance of having this dreadful disease. Phew. That's probably much lower than you thought.

This surprising result is an effect of the low base rate 1%. Such a small number of people actually have the disease that the false positives are relatively more numerous. This is why, if you test positive for some rare disease, you shouldn't necessarily freak out. Just get retested, while you explain the relevant principles of probability to your physician. You almost certainly understand them better.

## THE LANGUAGE OF PROBABILITY

| | Positive | Negative | Total |
|---|---|---|---|
| Sick | 98 | 2 | 100 |
| Okay | 297 | 9,603 | 9,900 |
| Total | 395 | 9,605 | 10,000 |

Sick, Okay, Pos, and Neg are known as ***events***. Even if you think of being Sick as a state, *picking* someone Sick out of a group is an event.

Sick is the opposite of Okay. The technical term is ***complement***, because together, the two events cover everybody.

$$\text{Sick} = \text{NOT Okay} = {\sim}\text{Okay} = \overline{\text{Okay}}$$

Those last two symbols for NOT (the squiggle and the bar over the top) are kind of evil. It's especially annoying that the bar is used over variables (like $\bar{x}$) to mean "***average***," but over events (like Okay) to mean "not." And the squiggle doesn't mean "approximately" here—it means "not." Huh? Yeah.

Since Sick and Okay cover everybody and they don't overlap as labels —no one is ever both Sick and Okay in this world—this set of events {Sick, Okay} is said to be ***MECE*** ("mee-see"):

- *Mutually **E**xclusive*: no one is both Sick and Okay.
- *Collectively **E**xhaustive*: everyone is either Sick or Okay.

{Pos, Neg} is also MECE: there are no gaps and no overlaps. Consultants love to throw around this acronym. While it's good to break down important issues in a MECE way, it's also good not to sound like a jackass, so use the term sparingly in civilian conversation.

Looking again at the grid, you can read off some probabilities.

| | **Positive** | **Negative** | **Total** |
|---|---|---|---|
| **Sick** | 98 | 2 | 100 |
| **Okay** | 297 | 9,603 | 9,900 |
| **Total** | 395 | 9,605 | 10,000 |

$P$(Neg) = "The probability of getting a negative test result."

$$= \frac{9{,}605}{10{,}000} = 0.9605$$

$$= 96.05\%$$

$P$(~Sick) = "The probability of picking someone who's *not* Sick out of the whole group."

$$= \frac{9{,}900}{10{,}000} = 0.99$$

$$= 99\%$$

## AND AND OR

What's the probability of picking someone who is both Okay and Negative? Find the box at the intersection of Okay and Negative:

| | **Positive** | **Negative** | **Total** |
|---|---|---|---|
| **Sick** | | | |
| **Okay** | | 9,603 | |
| **Total** | | | |

There are 9,603 people who are both Okay and Negative.

Then divide by the total population (10,000).

$$P(\text{Okay AND Neg}) = \frac{9{,}603}{10{,}000} = 0.9603 = 96.03\%$$

This can also be written as $P(\text{Okay} > \text{Neg})$. The > symbol is called an ***intersection***, and $P(\text{Okay} > \text{Neg})$ is the ***joint probability***. Each of the four boxes in the original 2×2 table represents an intersection.

| | **Positive** | **Negative** |
|---|---|---|
| **Sick** | Sick > Pos | Sick > Neg |
| **Okay** | Okay > Pos | Okay > Neg |

Notice that the number of people who are Okay AND Negative is less than the total number of people who are Okay. It's also less than the total number of people who are Negative.

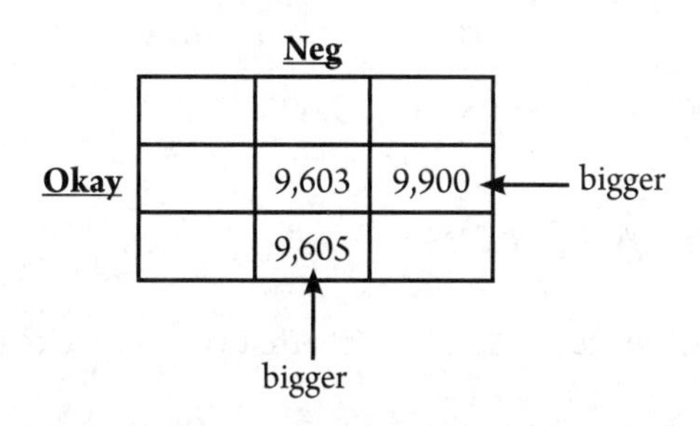


*AND* does not mean addition. You're applying the AND to the *condition* (you want people who are both Okay AND Negative), making the condition more stringent. This restricts the number of people who meet an AND condition. Think of it this way: how many folks have won the lottery? How many folks have been hit by lightning? Okay, how many folks have both won the lottery AND been hit by lightning? The answer to the last question is the smallest number.

In contrast, if you want to find people who are Okay ***OR*** Negative, you're relaxing the condition. People can be either Okay OR Negative (or both), so three of the four boxes in the 2×2 meet the condition:

| | **Pos** | **Neg** |
|---|---|---|
| **Sick** | ☒ | 2 |
| **Okay** | 297 | 9,603 |

$\longrightarrow \quad 2 + 297 + 9{,}603 = 9{,}902$

Only the 98 people who are both Sick and Positive are left out.

$$P(\text{Okay OR Neg}) = \frac{9{,}902}{10{,}000} = 0.9902 = 99.02\%$$

This can also be written as *P*(Okay < Neg). The < symbol is called a ***union***.

You might know that you can add probabilities to get an OR probability. If you do this addition, remember to subtract off the overlap, or you'll end up double-counting it.

Okay OR Neg = All Okay + All Neg – Okay AND Neg

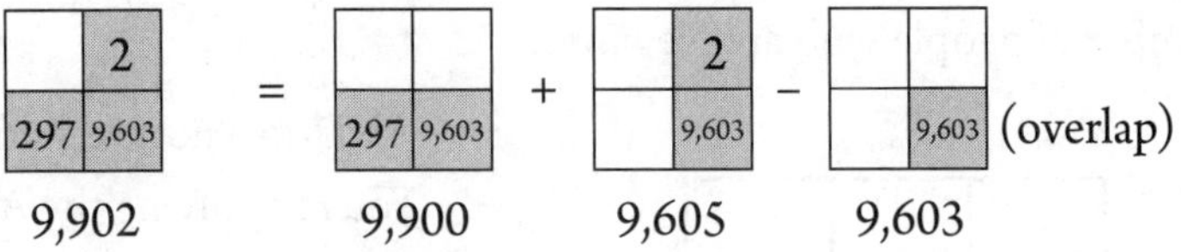

| | |
|---|---|
| | 2 |
| 297 | 9,603 |

9,902

=

| | |
|---|---|
| | |
| 297 | 9,603 |

9,900

+

| | |
|---|---|
| | 2 |
| | 9,603 |

9,605

–

| | |
|---|---|
| | |
| | 9,603 |

(overlap)

9,603

## CONDITIONAL PROBABILITY

What is the probability of a positive test result GIVEN that you are Sick?

In other words, you're not just 1 of the 10,000 anymore. You're 1 of the 100 sick people.

| | Positive | Negative | Total |
|---|---|---|---|
| **Sick** | | | 100 |
| **Okay** | | | ~~9,900~~ |
| **Total** | | | ~~10,000~~ |

Given this *condition*, what's the chance you get a positive test result?

| | Positive | Negative | Total |
|---|---|---|---|
| **Sick** | 98 | 2 | 100 |
| **Okay** | ~~X~~ | ~~X~~ | ~~9,900~~ |
| **Total** | ~~X~~ | ~~X~~ | ~~10,000~~ |

You're in the Sick row. 98 out of 100 sick people test positive. So the probability we want is $\frac{98}{100} = 0.98 = 98\%$.

This kind of probability is called ***conditional***. It's written with a vertical bar that means "given":

$P(\text{Pos} \mid \text{Sick}) =$ "the probability of testing positive, GIVEN that you're Sick"

↑ "given"

Notice the difference:

| | **Positive** | **Negative** | **Total** |
|---|---|---|---|
| **Sick** | 98 | | 100 |
| **Okay** | | | |
| **Total** | | | |

$$P(\text{Pos} \mid \text{Sick}) = \frac{98}{100}$$

| | **Positive** | **Negative** | **Total** |
|---|---|---|---|
| **Sick** | 98 | | |
| **Okay** | | | |
| **Total** | | | 10,000 |

$$P(\text{Pos} \cap \text{Sick}) = \frac{98}{10{,}000}$$

All that's different is what you divide the 98 by—but that makes all the difference. You can relate the probabilities to each other:

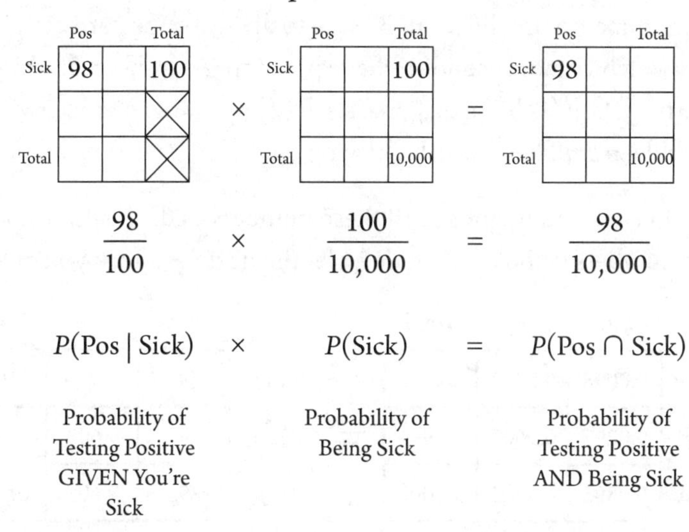

This formula can be rearranged:

$$P(\text{Pos} \mid \text{Sick}) = \frac{P(\text{Pos} \cap \text{Sick})}{P(\text{Sick})}$$

Intersections are the same regardless of order:

$$P(\text{Pos} > \text{Sick}) = P(\text{Sick} > \text{Pos}) = \frac{98}{10{,}000}$$

In contrast, conditionals most definitely depend on order:

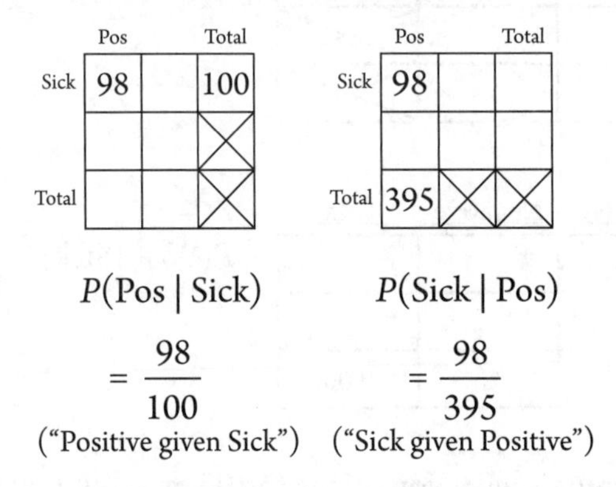


In each case, you use the 98 people who are both Sick and Positive, but the denominators are different. Be sure to divide by the correct "given" group, which comes second in the expression. As a shortcut, think of the vertical slash (|) like a diagonal slash (/) meaning "divided by." This is some deep stuff here, worth chewing on.

Again, in class you might see all these numbers as decimals. Switch to whole numbers if you can, but don't let the decimals freak you out.

| | **Pos** | **Neg** | **Total** |
|---|---|---|---|
| **Sick** | 0.0098 | 0.0002 | 0.01 |
| **Okay** | 0.0297 | 0.9603 | 0.99 |
| **Total** | 0.0395 | 0.9605 | 1.00 |

=

| | **Pos** | **Neg** | **Total** |
|---|---|---|---|
| **Sick** | 98 | 2 | 100 |
| **Okay** | 297 | 9,603 | 9,900 |
| **Total** | 395 | 9,605 | 10,000 |

## BAYES' RULE

The original big question—what's the probability that you're actually Sick, given a Positive test result—can now be written in probability-speak.

$$P(\text{Pos} \mid \text{Okay}) = 3\% \text{ (false positive rate)}$$

$$P(\text{Neg} \mid \text{Sick}) = 2\% \text{ (false negative rate)}$$

$$P(\text{Sick}) = 1\% \text{ (base rate of illness)}$$

What is $P(\text{Sick} \mid \text{Pos})$?

Bayes' Rule is a complicated formula used to compute the answer. We've chosen to omit this formula, because it's downright sadistic. Good news: you already computed the answer using the grid.

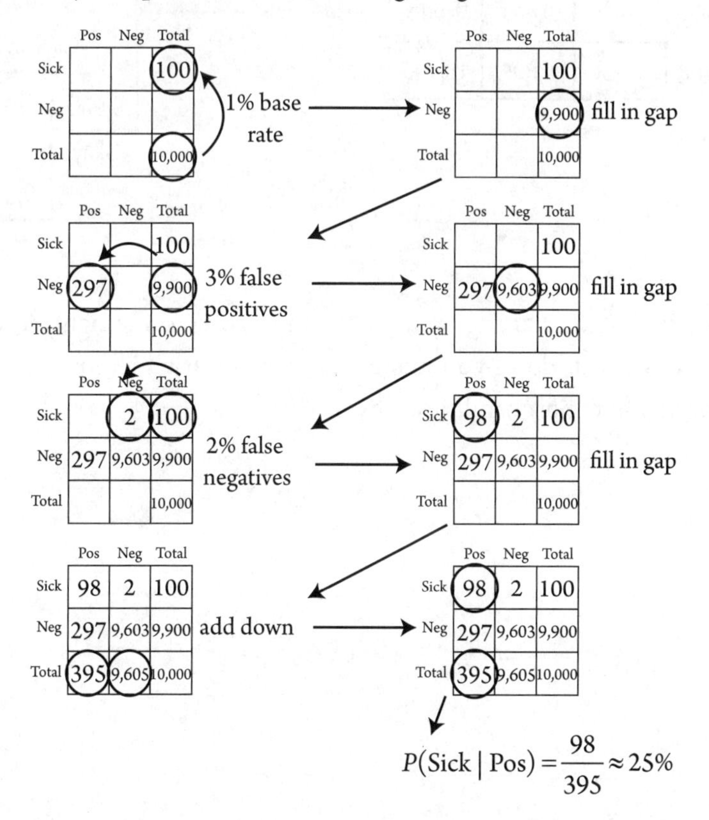

If you can, avoid the Bayes' Rule formula. Either use the grid above or a probability tree described below.

## PROBABILITY TREES

If you're not a fan of grids, you can represent the situation with a ***probability tree*** instead. As the tree branches out to the right, it subdivides the original group further and further.

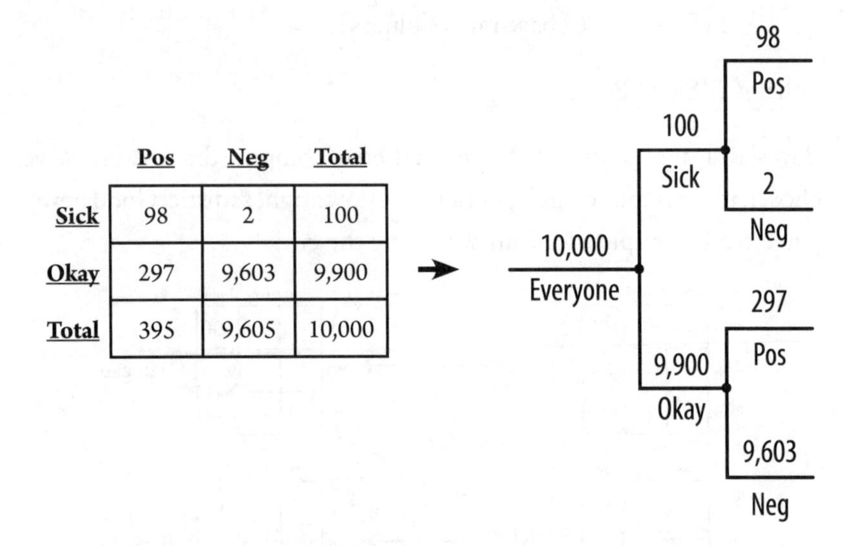

| | Pos | Neg | Total |
|---|---|---|---|
| Sick | 98 | 2 | 100 |
| Okay | 297 | 9,603 | 9,900 |
| Total | 395 | 9,605 | 10,000 |



You can imagine starting on the left and moving to the right. At each junction—each dot—you make a choice about which path to take according to the probabilities shown.

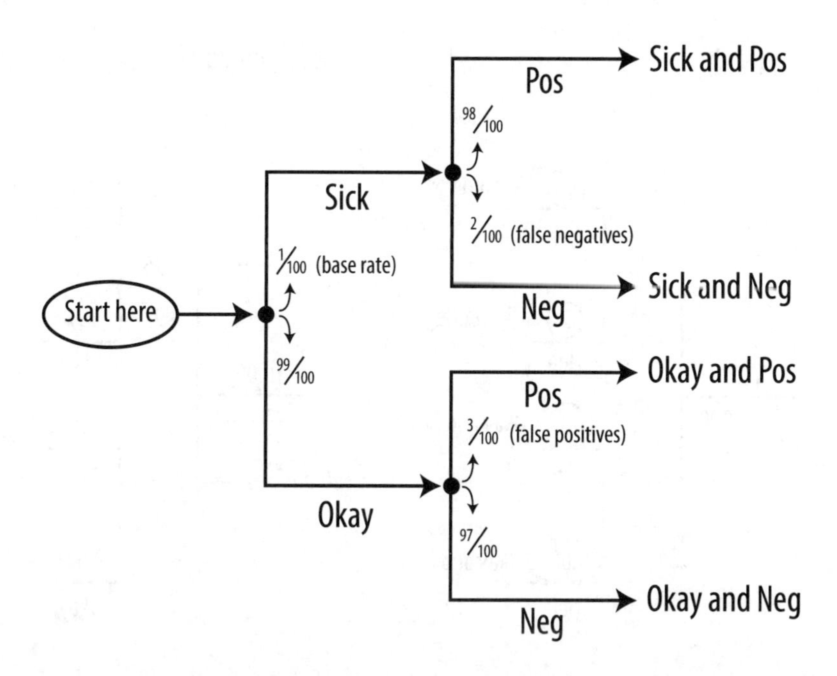


Make sure that the branches at each junction are MECE and that the probabilities at each junction add up to 1 (100%). In comparison to the grid, the big disadvantage of trees is that you only see *one* set of subtotals (e.g., the total number of Sick people = 100 and the total number of Okay people = 9,900). However, you need the subtotals of Positive and Negative test results as well. The key to doing Bayes' Rule with trees is to set up the tree one way, fill in all the numbers, then rearrange to get the other subtotals:

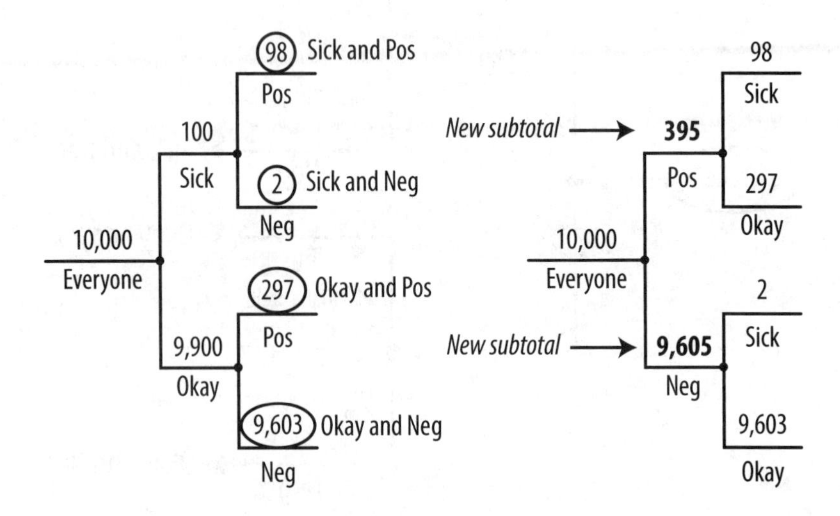


Notice how the four numbers on the ends of the tree (the "leaves") are the same as before, but they've been shuffled to allow the middle tier of the tree to show the positive/negative subtotals (395 and 9,605). The 395 is necessary to figure out $P(\text{Sick} \mid \text{Pos}) = \frac{98}{395} \approx 25\%$.

Long story short: you can do Bayes' Rule with a tree, but be ready to draw two of them. The grid is easier overall, but b-school has a thing for trees, so you should become familiar with them.

## INDEPENDENCE DAY

The last probability topic is ***independence*** (along with its opposite, ***dependence***). The question is:

> Does knowing whether one outcome happened *change your expectations* regarding another outcome?

If so, then the two outcomes are called ***dependent***. Careful: this doesn't mean that one outcome literally *depends* on the other. All it means is that knowing about one of them *influences your thinking* about the other one. If you know whether A happened, would you change your bet about whether B happened? If yes, then A and B are dependent. If not, then A and B are ***independent***.

A classic case of independence is a coin toss and a roll of a die. Frank secretly flips a coin and rolls a die. Before Frank shows you the coin, he shows you the die: it's a 3. Would you change your bet on the coin? Of course not. That's independence.

In contrast, the medical test is deliberately NOT independent of the disease. If you test positive for the disease, your expectations about truly having the disease rise from the 1% base rate to roughly 25%. $P(\text{Sick})$ and $P(\text{Sick} \mid \text{Pos})$ are not at all the same. That's dependence.

With a coin flip and a roll of the die, you have this grid (filled in with 120 as the smart number of total experiments):

| | | Die | | | | | | |
|---|---|---|---|---|---|---|---|---|
| | | 1 | 2 | 3 | 4 | 5 | 6 | Total |
| Coin | Heads | 10 | 10 | 10 | 10 | 10 | 10 | 60 |
| | Tails | 10 | 10 | 10 | 10 | 10 | 10 | 60 |
| | Total | 20 | 20 | 20 | 20 | 20 | 20 | 120 |

$$P(\text{Heads} \mid 3) = \text{“probability of flipping Heads, given that you rolled a 3”} = 10/20 = 1/2$$

$$P(\text{Heads}) = 60/120 = 1/2$$

Since these two probabilities are the same, the outcome "Heads" and the outcome "3" are independent.

The other test for independence is to see whether the joint probability—the chance that both outcomes occur—equals the product of each separate unconditional probability.

$$\underset{\text{flip heads and roll a 3}}{P(\text{Heads AND } 3)} \stackrel{?}{=} \underset{\text{flip heads}}{P(\text{Heads})} \cdot \underset{\text{roll a 3}}{P(3)}$$

$$\frac{10}{120} \stackrel{?}{=} \frac{1}{2} \cdot \frac{1}{6}$$

$$\frac{1}{12} \underset{\text{true}}{\stackrel{\checkmark}{=}} \frac{1}{2} \cdot \frac{1}{6}$$

Since this is true, Heads and 3 are independent. In contrast, Sick and Positive are *not* independent (they are dependent), because this equation doesn't hold.

| | **Pos** | **Neg** | **Total** |
|---|---|---|---|
| **Sick** | 98 | 2 | 100 |
| **Okay** | 297 | 9,603 | 9,900 |
| **Total** | 395 | 9,605 | 10,000 |

$$P(\text{Sick AND Pos}) \stackrel{?}{=} P(\text{Sick}) \cdot P(\text{Pos})$$

$$\frac{98}{10{,}000} \stackrel{?}{=} \frac{100}{10{,}000} \cdot \frac{395}{10{,}000}$$

$$\frac{98}{10{,}000} \underset{\text{Not true}}{\stackrel{?}{=}} \frac{395}{1{,}000{,}000}$$

## DECISION TREES

You can represent probabilities with trees:

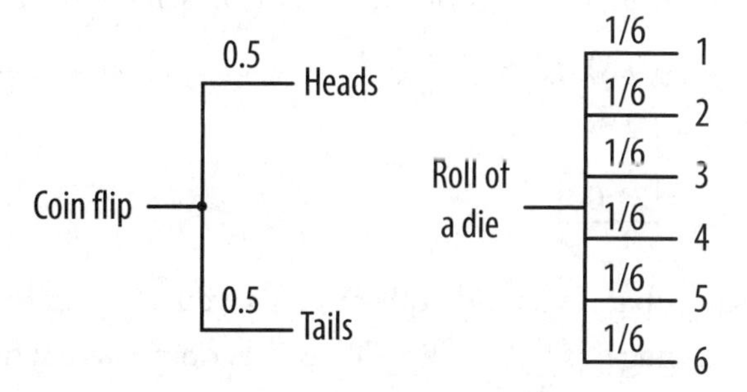


Label each branch with its probability.

You can also represent payoffs on each branch and calculate the expected value of a tree, which would be the average payoff if you played the game a lot of times.

Game: Heads = you win \$5. Tails = you lose \$3.

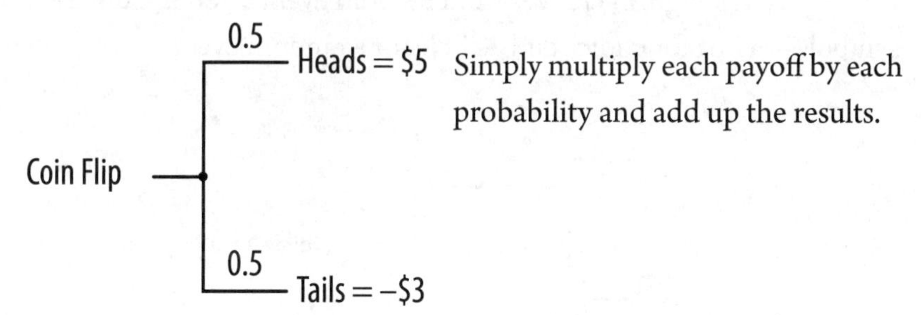


Simply multiply each payoff by each probability and add up the results.

Expected value of a coin flip:

$$
\begin{aligned}
E(\text{flip}) &= \begin{pmatrix}\text{Chance}\\ \text{of Heads}\end{pmatrix}\begin{pmatrix}\text{Payoff}\\ \text{of Heads}\end{pmatrix} + \begin{pmatrix}\text{Chance}\\ \text{of Tails}\end{pmatrix}\begin{pmatrix}\text{Payoff}\\ \text{of Tails}\end{pmatrix}\\
&= (0.5)(\$5) + (0.5)(-\$3)\\
&= \$2.50 - \$1.50\\
&= \boxed{\$1.00}
\end{aligned}
$$

This $1 means that if you did 100 coin flips, you'd expect to earn $100 total, or an average of $1 per flip. Of course, on no actual flip can you earn precisely $1, but that's what you'd expect as a long-run average, so if your pockets were deep enough to handle occasional runs of $3 losses, you'd want to play this game over and over. Like a casino, you'd come out ahead in the end.

Finally, you can represent choices or ***decisions*** with trees. For instance, you might have a choice between the coin flip game and receiving $0.90 for sure. To distinguish choices from random events, you use different symbols—a box for choice and a circle for a random event.

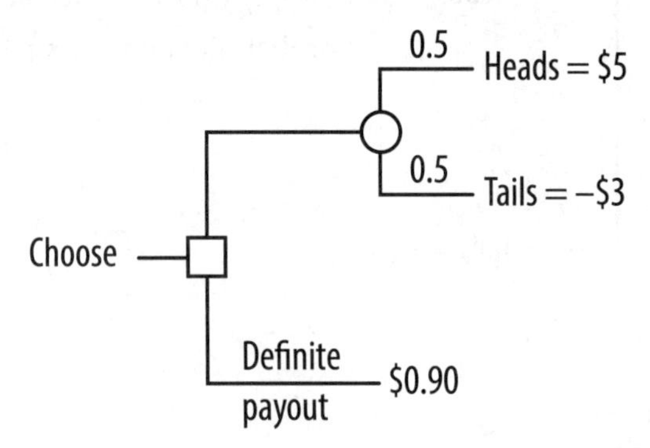


The expected value of the coin flip is $1, so if you can stand the risk of losing $3 or if you can play the game over and over (taking the risk of losses along the way)—you'd probably choose coin flip. But if the numbers get much larger, armchair psychology says that you'll become more ***risk-averse*** and will likely choose the certain payout.

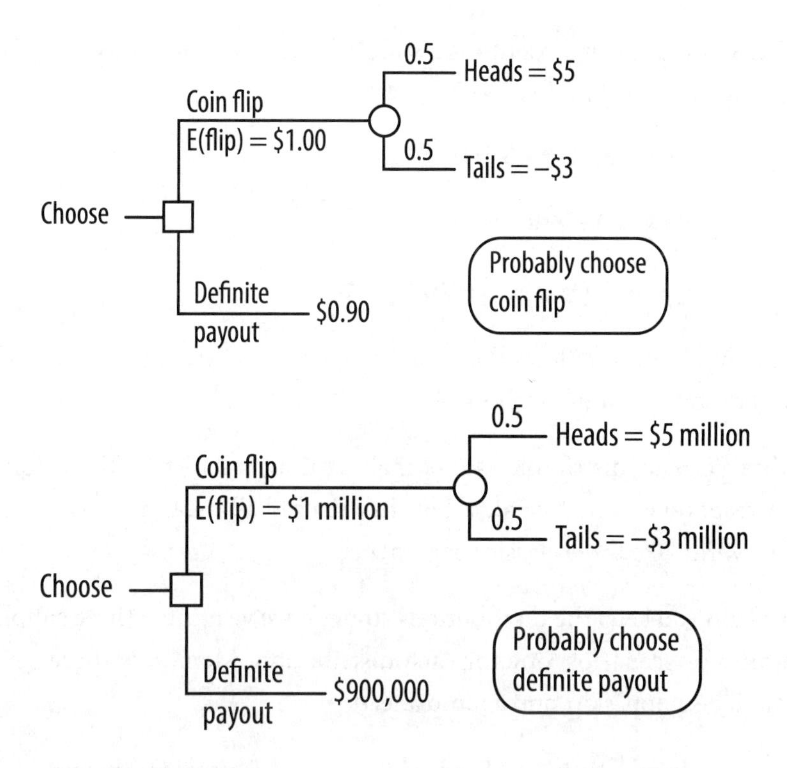


## DISTRIBUTIONS

***Distributions*** are essentially histograms.

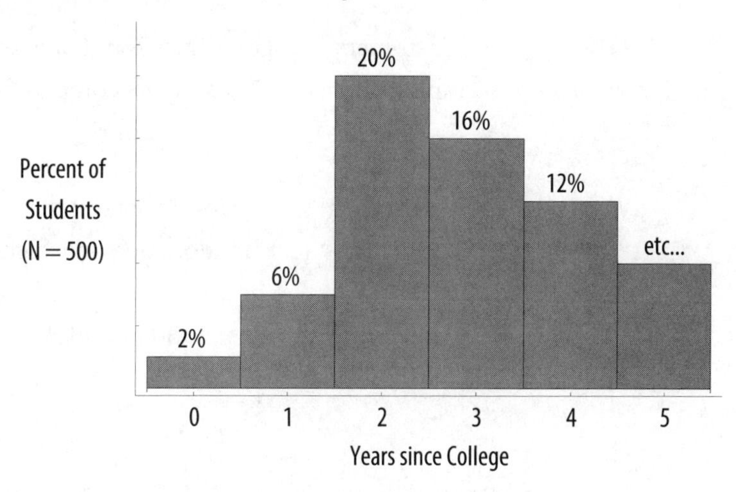

You use percents in order to show probabilities. The previous distribution is equivalent to this list:

$$P(0 \text{ years}) = 2\% = 0.02$$

$$P(1 \text{ year}) = 6\% = 0.06$$

$$P(2 \text{ years}) = 20\% = 0.20$$

The distribution reflects the actual frequency of "years since college" you observe in your b-school class.

Three "classic" distributions appear all the time in statistics. These don't correspond exactly to reality, but they're used all the time to simulate the world, like three classic soap operas.

To help you keep the distributions straight, we've created three simple stories or scenarios, one for each distribution. Memorize these, and you'll be a hop, skip, and a jump ahead.

| | Distribution | Scenario | Typical Question |
|---|---|---|---|
| 1. | ***Binomial*** | A bunch of coin flips | How likely am I to get exactly 6 heads in 10 flips? |
| 2. | ***Poisson*** ("pwah-son") | Customers coming into a store | How likely is it that exactly 7 customers come in this hour? |
| 3. | ***Normal*** | Weights in a population | How likely is it that a random person weighs between 100 pounds and 130 pounds? |

## THE BINOMIAL DISTRIBUTION

This scenario involves a bunch of coin flips.

- You flip a coin a certain number of times. The number of flips is $n$. Each flip is independent of every other flip.
- Each flip has two possible outcomes:

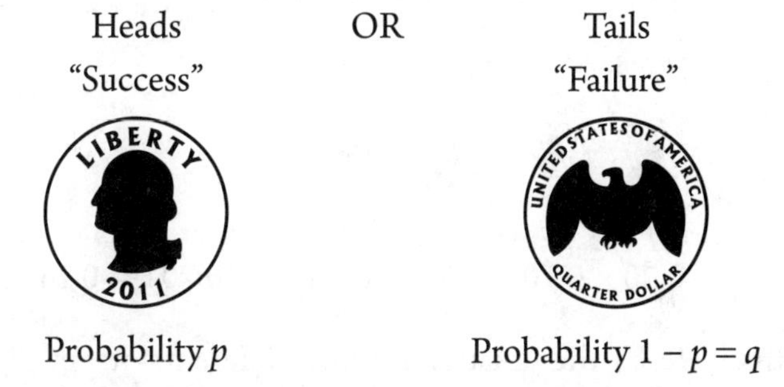


Lowercase $p$ is always used for the probability of success. The probability of failure is $q = 1 - p$.

- The coin doesn't have to be fair, although it could be.

| | |
|---|---|
| $P(\text{Heads}) = p = 0.50$ | Fair coin |
| $P(\text{Heads}) = p = 0.75$ | Heads more probable |
| $P(\text{Heads}) = p = 0.30$ | Tails more probable |

- Whatever $p$ is, it stays the same for all flips. Use the same coin throughout.
- The binomial distribution tells you the probability of any particular total number of Heads.

Say you flip a weighted coin four times. The coin is weighted so that on any given flip, the probability of Heads is 70%. The binomial distribution fills in the blanks below:

| | | | | |
|---|---|---|---|---|
| Chance of | 4 Heads | in 4 flips | $= P(x = 4)$ | = _______ |
| " | 3 Heads | in 4 flips | $= P(x = 3)$ | = _______ |

| | | | | |
|---|---|---|---|---|
| " | 2 Heads | in 4 flips | $= P(x=2)$ | = ______ |
| " | 1 Heads | in 4 flips | $= P(x=1)$ | = ______ |
| " | 0 Heads | in 4 flips | $= P(x=0)$ | = ______ |

You know enough from the GMAT or GRE to fill in the first blank. Since each flip is independent, and $P(\text{Heads}) = 0.70 = 70\%$, then the probability of four Heads in a row is this:

4 Heads = Heads Heads Heads Heads

$$p \times p \times p \times p = p^4$$
$$0.70 \times 0.70 \times 0.70 \times 0.70 = (0.70)^4 = 0.2401 \approx 24\%$$

For three Heads, the situation is a little different. First of all, you've got one Tails mixed in, since you've got four flips and no other options besides Heads and Tails.

3 Heads = Heads Heads Heads Tails

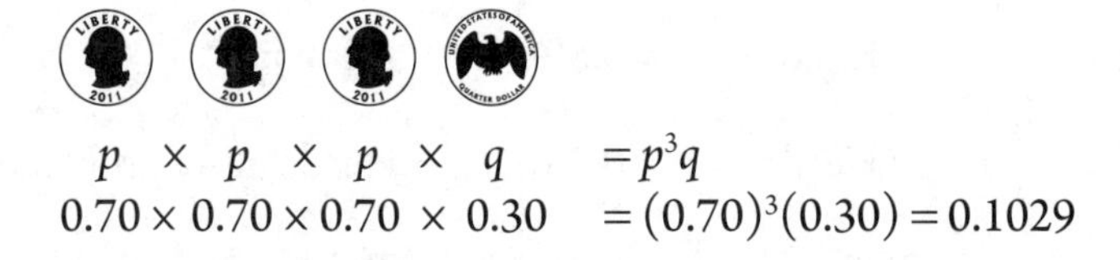

$$p \times p \times p \times q = p^3q$$
$$0.70 \times 0.70 \times 0.70 \times 0.30 = (0.70)^3(0.30) = 0.1029$$

The other complication is that there's more than one sequence of flips that gives you exactly three Heads and one Tails. It turns out there are four such sequences:

| | | | | | |
|---|---|---|---|---|---|
| 1. | H | H | H | T | 4 different ways |
| 2. | H | H | T | H | |
| 3. | H | T | H | H | |
| 4. | T | H | H | H | |

If you remember the combinations formula from the GMAT or GRE, here it is again, rearing its ugly head...

$$\frac{\text{Flips!}}{\text{Heads! Tails!}} = \frac{4!}{3!\ 1!} = 4 \text{ different ways}$$

Those exclamation points are ***factorials***. The value of 4-factorial (written as "4!") is $4 \times 3 \times 2 \times 1 = 24$. The value of 3-factorial is $3 \times 2 \times 1 = 6$. All in, the probability of three Heads (and one Tails) is this:

| Probability of 1 arrangement | × | Number of arrangements | = | $P$(3 Heads) |
|---|---|---|---|---|
| HHHT | | $\frac{\text{Flips!}}{\text{Heads! Tails!}}$ | | |
| $p \cdot p \cdot p \cdot q$ | × | $\frac{4!}{3!\ 1!}$ | = | $4p^3q$ |
| (0.70)(0.70)(0.70)(0.30) | × | 4 | = | 0.4116 ≈ 41% chance |

For two heads (and two tails), the same calculation works out this way:

| Probability of 1 arrangement | × | Number of arrangements | = | $P$(2 heads) |
|---|---|---|---|---|
| HHTT | | $\frac{\text{Flips!}}{\text{Heads! Tails!}}$ | | |
| $p \cdot p \cdot q \cdot q$ | × | $\frac{4!}{2!\ 2!}$ | = | $6p^2q^2$ |
| (0.70)(0.70)(0.30)(0.30) | × | 6 | = | 0.2646 ≈ 26% chance |

One Heads and three Tails:

| Probability of 1 arrangement | × | Number of arrangements | = | $P(1 \text{ heads})$ |
|---|---|---|---|---|
| H T T T | | $\frac{\text{Flips!}}{\text{Heads! Tails!}}$ | | |
| $p \cdot q \cdot q \cdot q$ | × | $\frac{4!}{1!\ 3!}$ | = | $4pq^3$ |
| $(0.70)(0.30)(0.30)(0.30)$ | × | 4 | = | 0.0756 ≈ 7.6% chance |

Zero Heads and four Tails:

| Probability of 1 arrangement | × | Number of arrangements | = | $P(0 \text{ heads})$ |
|---|---|---|---|---|
| T T T T | | $\frac{\text{Flips!}}{\text{Heads! Tails!}}$ | | |
| $q \cdot q \cdot q \cdot q$ | × | $\frac{4!}{0!\ 4!}$ | = | $4pq^3$ |
| $(0.30)(0.30)(0.30)(0.30)$ | × | 1 | = | 0.0081 ≈ 0.8% chance |

By the way, 0-factorial, or 0!, equals 1.

The graph of all these outcomes looks like this:

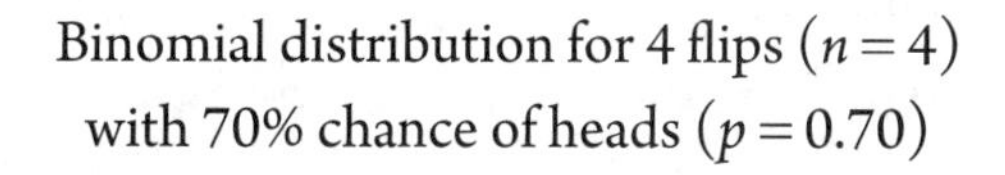


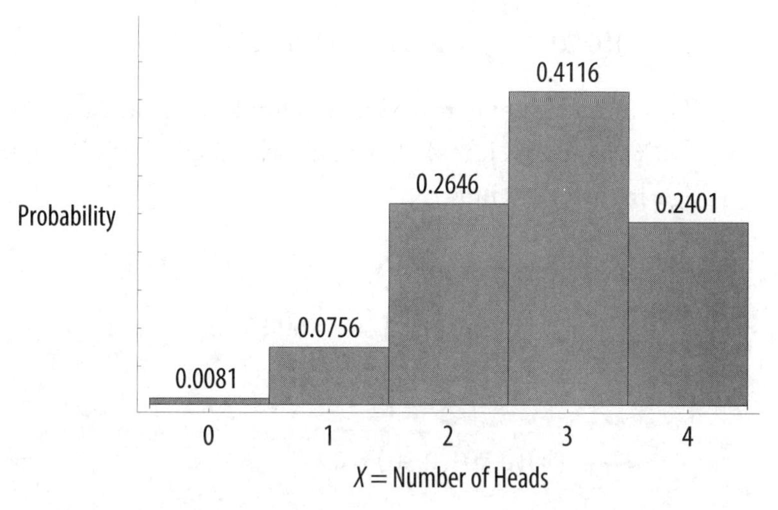


The binomial distribution depends on the total number of flips ($n = 4$ in this case) and the weighting of the coin ($p = 0.70$ in this case). For other values of $n$ and $p$, you'll get a different set of probabilities.

The general formula for the binomial distribution with a given $n$ and $p$ (and with $q = 1 - p$) is:

$$P(x = \text{ number of Heads}) = \frac{n!}{x!(n-x)!} p^x q^{n-x}$$

Compare to the earlier case.

$n = 4$
$p = 0.70$
$q = 0.30$

| | | | | |
|---|---|---|---|---|
| $P(x=4)$ | = | 1 | $p^4$ | $= 0.2401$ |
| $P(x=3)$ | = | 4 | $p^3q$ | $= 0.4116$ |
| $P(x=2)$ | = | 6 | $p^2q^2$ | $= 0.2646$ |
| $P(x=1)$ | = | 4 | $pq^3$ | $= 0.0756$ |
| $P(x=0)$ | = | 1 | $q^4$ | $= 0.0081$ |

The mean, variance, and standard deviation of the binomial distribution have simple formulas.

Mean $\mu = np$

$$\mu = (4)(0.70) = 2.8 = \text{mean number of Heads}$$

This formula should make sense. If $n = 100$ flips and $p = 0.60$ (60% chance of Heads), then you'd expect 60 Heads, on average, in those 100 flips.

$$np = \mu$$

$$(100)(0.60) = 60 \text{ Heads}$$

Variance $\sigma^2 = npq = np(1 - p)$

$$\sigma^2 = \longleftarrow (4)(0.70)(0.30) = 0.84$$

This formula just "is." Sorry.

Standard Deviation $\sigma = \sqrt{\sigma^2} = \sqrt{npq} = \sqrt{np(1-p)}$

$$\sigma = \sqrt{0.84} \approx 0.92$$

If $p$ is exactly 0.5, the binomial distribution is symmetrically hump-shaped:

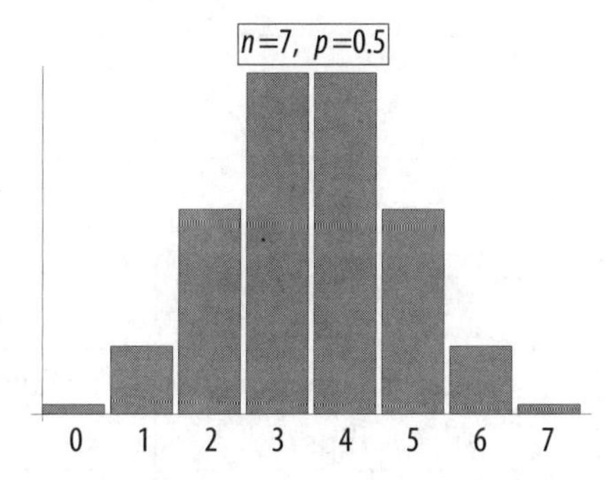


The closer $p$ is to 1 or to 0, the more skewed the distribution from one end to the other.

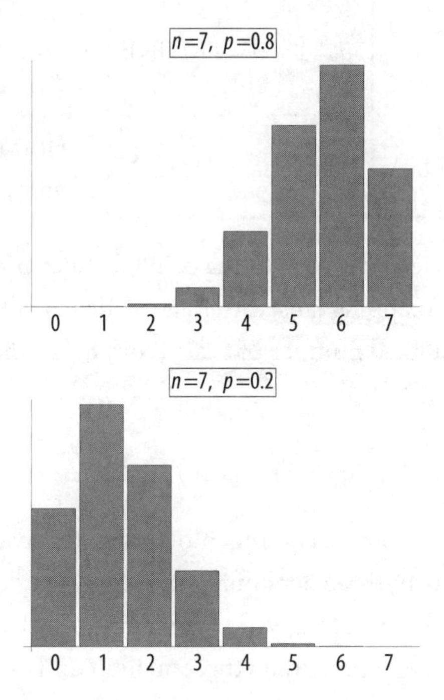


Since you'll encounter problems that aren't literal coin flips, you will need to draw analogies. To use the binomial distribution, you have to be able to map the given problem to *a fixed number of coin flips*. Here's an example. How many people in your company have January birthdays?

| Fixed number $n$ | Number of flips | Number of people in your company |
|---|---|---|
| Independent events | Flips | Birthdays |
| Success | Heads | Birthday in January |
| Probability $p$ of success | $p = P(\text{Heads})$ | $p = P(\text{Birthday in Jan}) \approx 1/12$ |
| Constant $p$? | $p$ stays same for all flips | $p$ stays same for all people |
| Failure | Tails | Birthday in another month |
| Probability $q$ of failure | $q = 1 - p = P(\text{Tails})$ | $q = P(\text{Birthday in another month}) \approx 11/12$ |
| Random variable $X$ | How many Heads? | How many people with Jan birthdays? |

Do this kind of clear mapping to the coin flips scenario, and you won't go wrong. If the mapping fails, the situation doesn't fit. For instance, if you don't have a fixed number of trials (coin flips), then you can't use the binomial distribution.

## THE POISSON DISTRIBUTION

The scenario is customers coming into a store. On average, they arrive at some set rate (say, three per hour—this store isn't doing so hot). But the customers also arrive totally independently of each other, at random, so sometimes you get more than three an hour and other times less:

Long-run average: three customers per hour

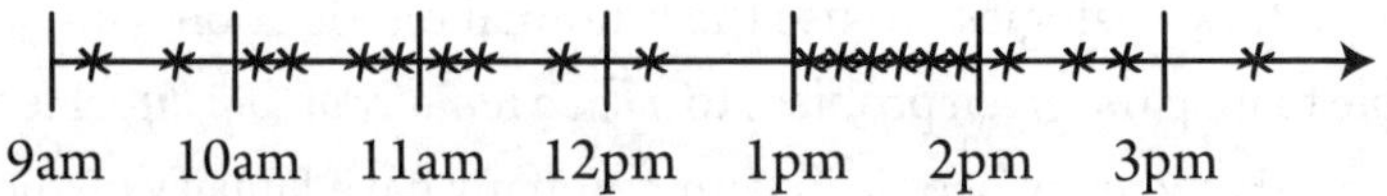

These customer arrivals can then be modeled as a so-called ***Poisson process***, and the number of customers you actually do get in any given hour obeys the ***Poisson distribution***.

The formula for this distribution is a little ugly. First, you need to know that the letter ***e*** in math is a special number around 2.72 that's almost as important as $\pi$. Go math! You might have known and loved *e* way back when as the base of the natural logarithm (which is not an a cappella algebra group). This number *e* can be raised to a power; you can do this with the $e^x$ button on your calculator or EXP() function in Excel. It's better than bad; it's good.

If your average arrival rate is 3 customers per hour, then the chance that $x$ customers actually arrive in a particular hour is given by $P(x) = \frac{e^{-3} \cdot 3^x}{x!}$.

Let's see how that works. The chance that you get exactly *zero* customers in that hour is:

$$P(x=0) = \frac{e^{-3} \cdot 3^0}{0!} = \frac{e^{-3} \cdot 1}{1} \approx 0.0498$$

Particular number of customers this hour

Chance of exactly zero customers ≈ 4.98%

Check out the probabilities for other particular numbers of customers arriving in that hour. Think about the shape of the distribution.

$$P(x=1) = \frac{e^{-3} \cdot 3^1}{1!} = \frac{e^{-3} \cdot 3}{1} \approx 0.1494$$

Chance of exactly 1 customer arriving ≈ 14.94%

$$P(x=2) = \frac{e^{-3} \cdot 3^2}{2!} = \frac{e^{-3} \cdot 9}{2} \approx 0.2240$$

Chance of exactly 2 customers arriving ≈ 22.40%

$$P(x=3)=\frac{e^{-3}\cdot 3^3}{3!}=\frac{e^{-3}\cdot 27}{6}\approx 0.2240 \text{ again}$$

Chance of exactly 3 customers arriving ≈ 22.40%

$$P(x=4)=\frac{e^{-3}\cdot 3^4}{4!}=\frac{e^{-3}\cdot 81}{24}\approx 0.1680$$

Chance of exactly 4 customers arriving ≈ 16.80%

$$P(x=5)=\frac{e^{-3}\cdot 3^5}{5!}=\frac{e^{-3}\cdot 243}{120}\approx 0.1008$$

Chance of exactly 5 customers arriving ≈ 10.08%

$$P(x=6)=\frac{e^{-3}\cdot 3^6}{6!}=\frac{e^{-3}\cdot 729}{720}\approx 0.0504$$

Chance of exactly 6 customers arriving ≈ 5.04%

$$P(x=7)=\frac{e^{-3}\cdot 3^7}{7!}=\frac{e^{-3}\cdot 2{,}187}{5{,}040}\approx 0.0216$$

Chance of exactly 7 customers arriving ≈ 2.16%

This distribution technically goes on forever, but the values get tinier and tinier. In other words, it's not *impossible* for 20 customers to show up in a single hour—it's just very, very unlikely (about 7 chances in 100 billion).

The graph looks like this:

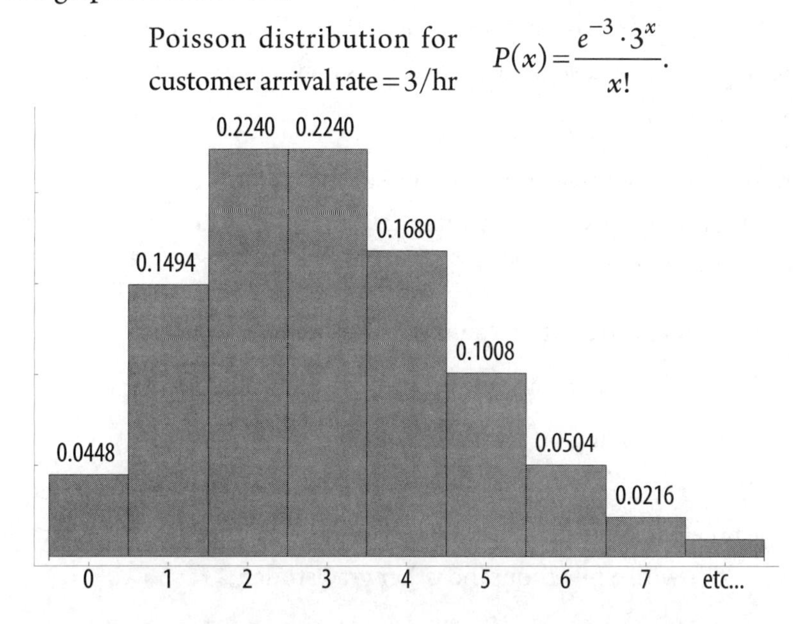


The average number of customers is three, as you'd expect—that was specified at the beginning. Magically, the variance also turns out to be three, so the standard deviation is $\sqrt{3}$. For Poisson distributions, the mean always equals the variance. How about that?

Mean $\mu = 3$ customers

Variance $\sigma^2 = 3$

Standard deviation $\sigma = \sqrt{3} \approx 1.7$ customers

In general, the arrival rate is given the Greek letter ***lambda***, which looks like $\lambda$. This is the most general formula:

Poisson distribution with a customer arrival rate $= \lambda$ per hour

$$P(x) = \frac{e^{-\lambda} \cdot \lambda^x}{x!}$$

Mean $\mu = \lambda$

Variance $\sigma^2 = \lambda$

Standard deviation $= \sqrt{\lambda}$

The Poisson distribution can substitute for the binomial distribution when $n$ is large (say, over 50) and $p$ is very small (say, less than 0.1). This can be useful because the binomial gets computationally annoying as $n$ gets larger and larger.

The big reason they teach you the Poisson distribution in business school is that lots of things a manager might want to keep track of are Poisson processes, or approximately so:

- Customers arriving at a store
- Jobs arriving at a computer server
- Errors occurring on an assembly line
- The slamming of porta-potty doors, when your RV is parked near a whole bank of them and you are camping out for Duke basketball tickets with a few thousand other grad students
- You get the idea.

## CONTINUOUS VS. DISCRETE

Pardon the interruption—a digression is necessary before we get to the third distribution (the so-called normal distribution). Up to now, the questions have all been about ***discrete random variables***—whole number outcomes.

- How many heads do I get in 10 coin flips, with $p = 0.6$ (60% chance of heads on each flip)?
- How many customers arrive this hour, if the average arrival rate is 3 customers per hour?

The possible answers are 0, 1, 2, 3, etc.—only whole numbers.

The distributions so far have told you how likely each of those outcomes is. Each outcome is discrete, so you draw histograms. Each discrete value on the $x$-axis has its own hash mark and column of probability.

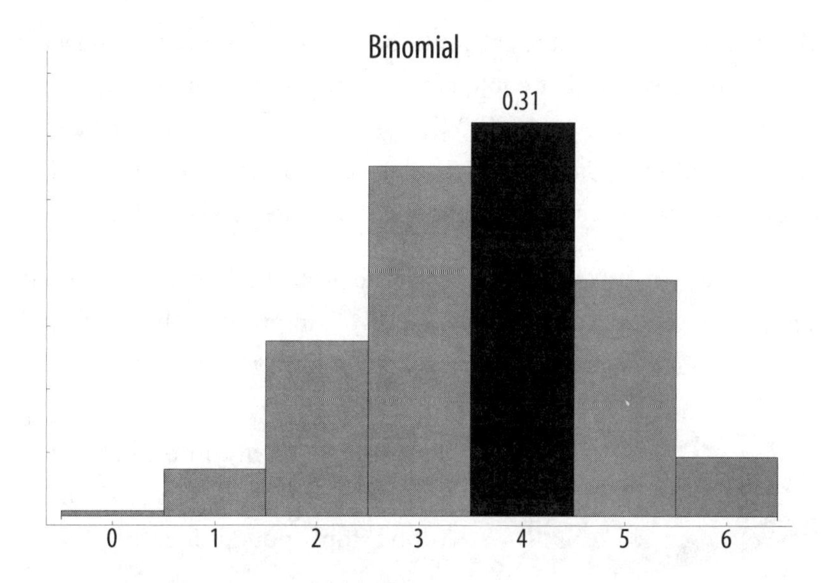


$P(x=4)=0.31$

The probability that $x=4$ is 31%.

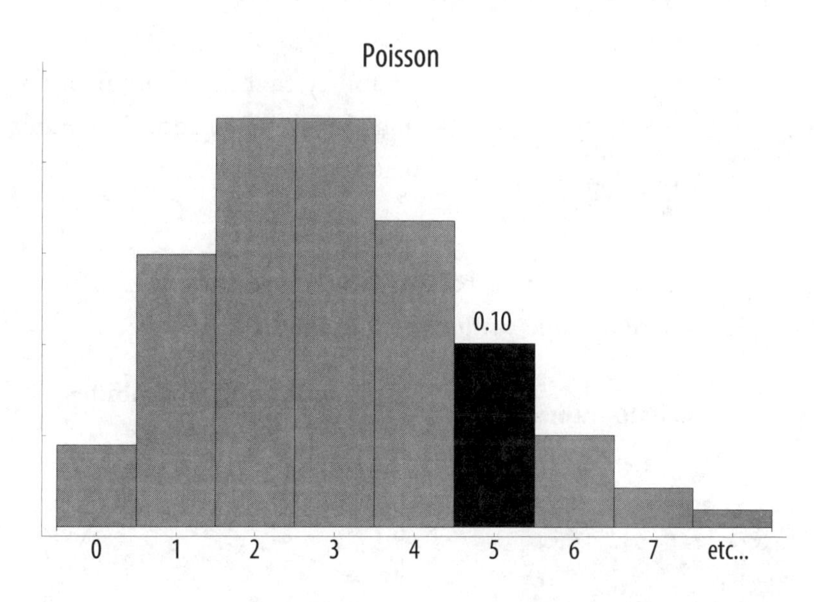


$P(x=5)=0.10$

The probability that $x=5$ is 10%.

Things change when there are continuous outcomes. What if you want to measure a weight that could take on *any* value, including ones with decimals, like 103.732 pounds? That kind of outcome is called a ***continuous random variable***—it can be any real number in some range. You assume you can measure the weight with infinite precision.

In this case, you no longer talk about the probability of *some particular number* as an outcome. Why? Because it's too improbable that someone's weight is exactly equal to some random, infinitely long decimal.

Probability that my brother-in-law weighs *exactly* 216.271381545992... (don't stop!) pounds is zero.

What you do instead is talk about a ***range*** of weights. You can sensibly measure the probability of a weight being in a range.

Probability that my brother-in-law weighs ***between*** 200 and 250 pounds is 0.36 (36%).

To represent this sort of probability, you use something *similar to* a histogram—a ***continuous probability distribution***.

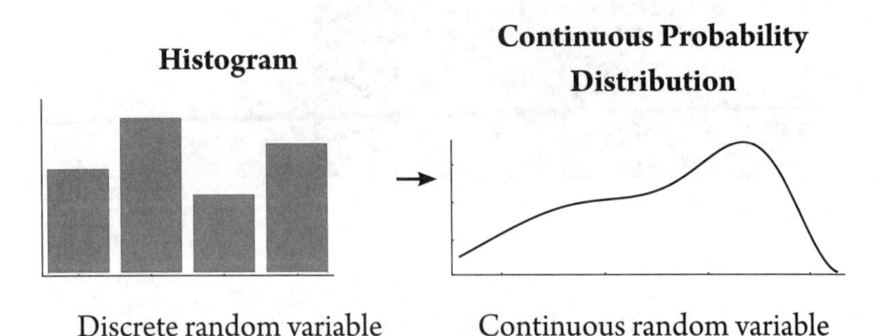

The mathematical difference is in the way you indicate a probability. For a discrete random variable, the *height* of the column is the probability itself:

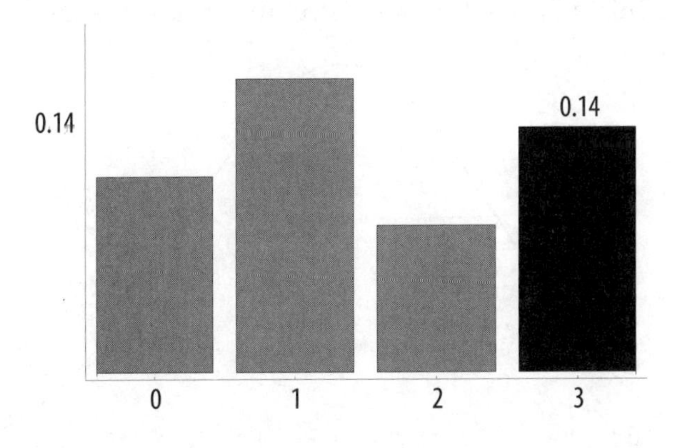


$$P(x = 3) = 0.14$$

The probability that $x$ is 3 is 14%.

On a continuous probability distribution, the *area under the curve* and between two end points is the probability.

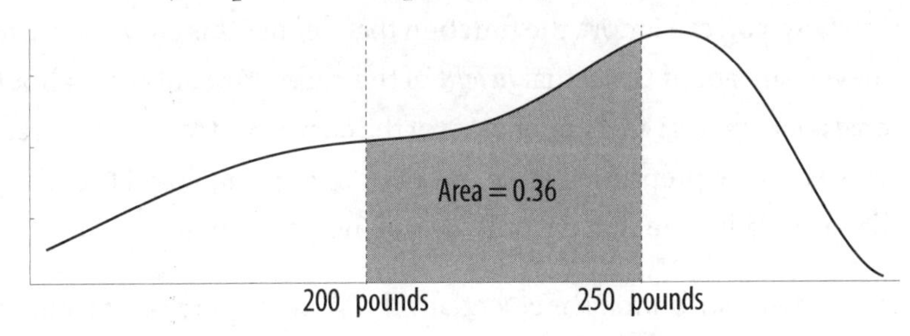


$$P(200 < x < 250) = 0.36$$

The probability that the weight is between 200 and 250 pounds is 36%.

The total area under any continuous distribution is 1 (100%), since the maximum probability is always 1 (nothing can be more than 100% likely to occur).

## THE NORMAL DISTRIBUTION

The most important distribution in all of statistics is the ***normal distribution***. This is the continuous distribution known as the ***bell curve***:

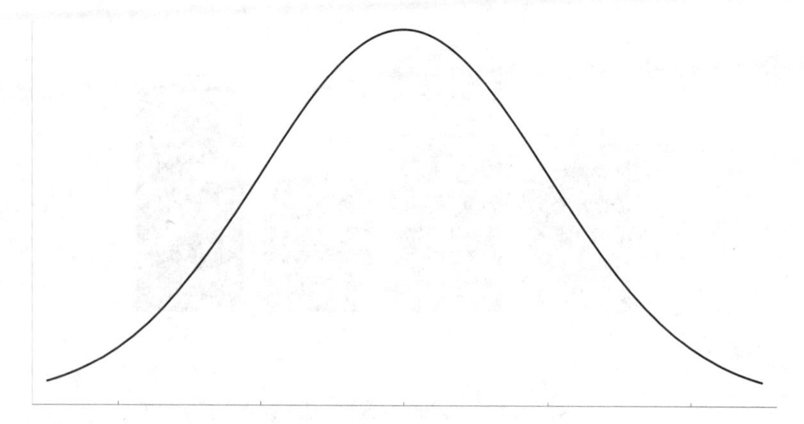

The curve never actually touches the *x*-axis, but it gets very close as it goes on forever in both directions.

Fortunately, you can ignore the function that defines this curve because you never care about the actual *height* of the curve. You only care about the *area under the curve,* or under parts of the curve. And to find that area, you have to use a preprinted table of values or a special Excel function, so the formula for the shape would do you no good anyway.

Every normal distribution has essentially the same shape: a central hump with two long, symmetrical tails on either side.

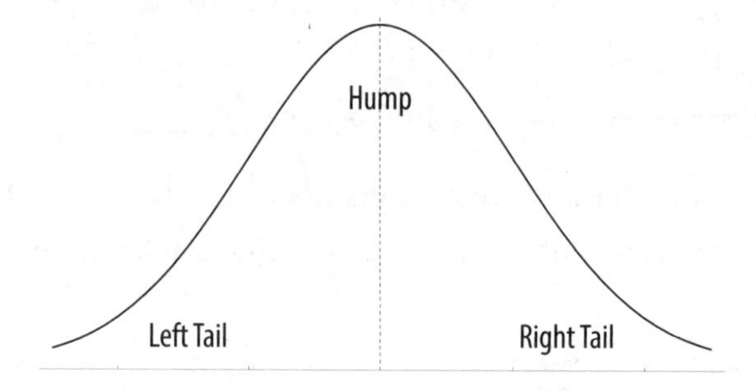

The peak of the hump is centered over the mean. So if some population of people has a mean weight of 150 pounds, and that weight is "normally distributed," then the distribution looks like this:

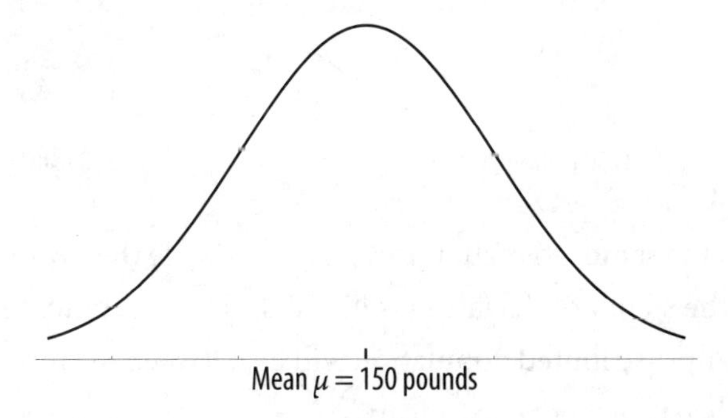


The curve is completely symmetrical around the mean.

The other thing you have to specify about a normal distribution is how spread out it is.

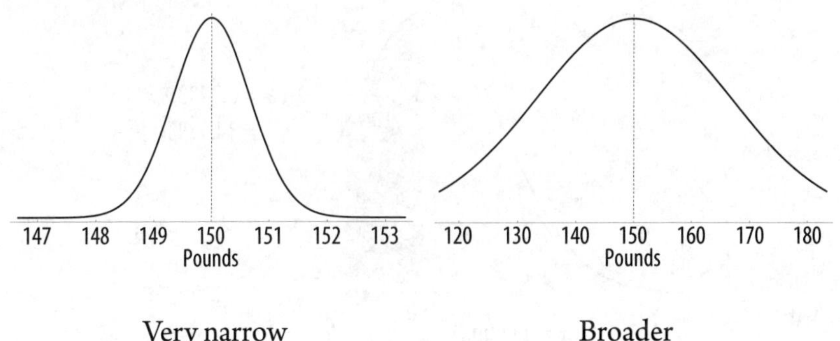


Very narrow
Small standard deviation

Broader
Bigger standard deviation

The standard deviation $\sigma$ provides a great yardstick for these normal curves. The area under the curve from the mean to a point exactly 1 standard deviation away from the mean is always 0.3413 (~34%).

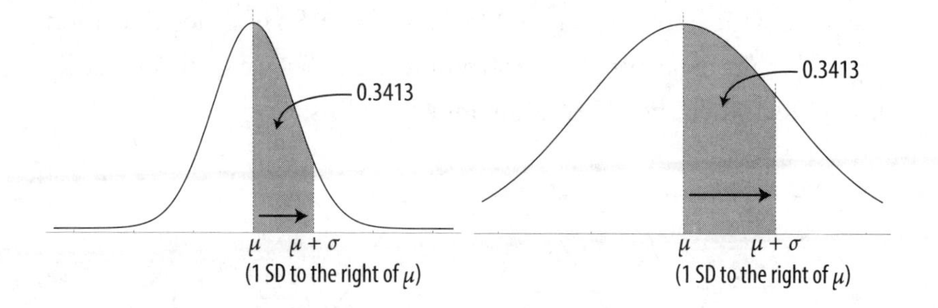


The area is the same, no matter how skinny or wide the normal curve is (how big the standard deviation is in units). In probability terms, 34% of a normally distributed population will lie between the mean and one standard deviation above the mean.

For instance, if the mean $\mu = 150$ pounds and the standard deviation $\sigma = 1$ pound, then 34.13% of the population weighs between 150 and 151 pounds:

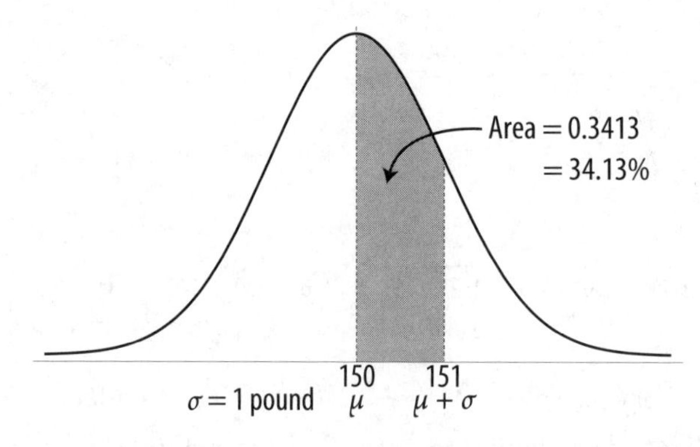


If the standard deviation $\sigma$ is instead 12 pounds, then 34.13% of the population weighs between 150 and 162 pounds:

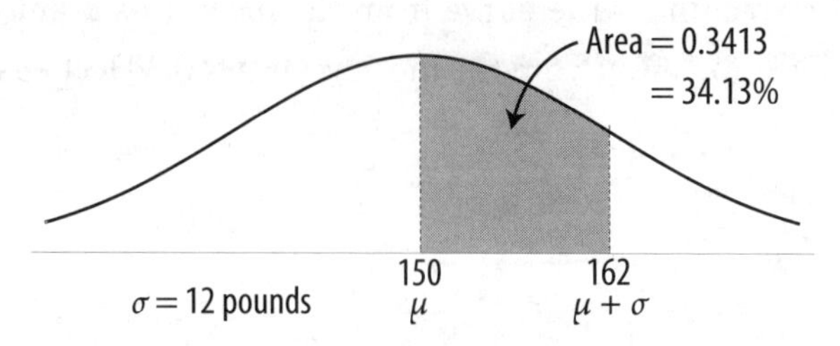

For the normal distribution, the following guidelines are always true. ~68% of the population is within 1 standard deviation of the mean.

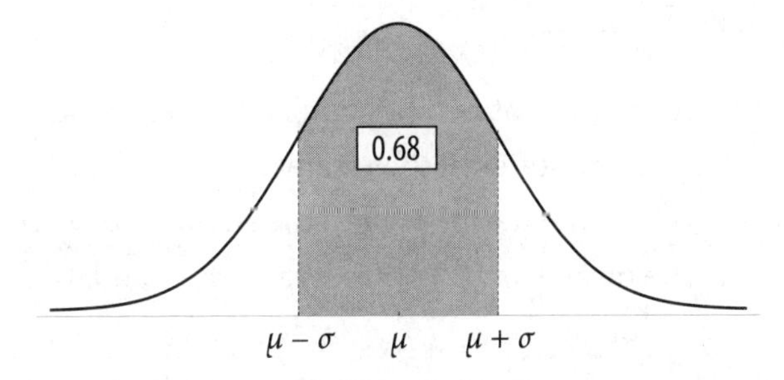


~95% of the population is within 2 standard deviations of the mean.

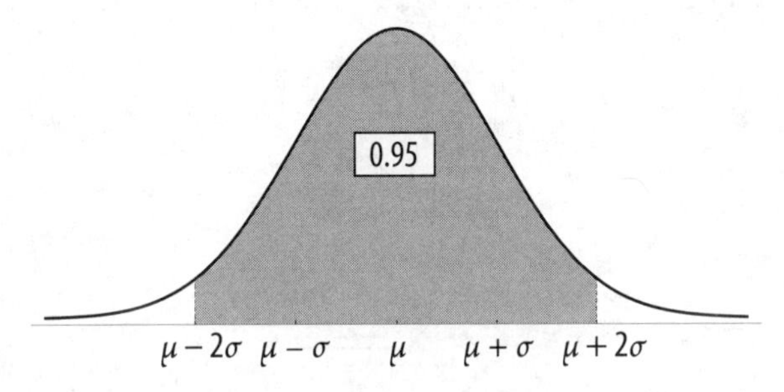


~99.7% of the population is within 3 standard deviations of the mean.

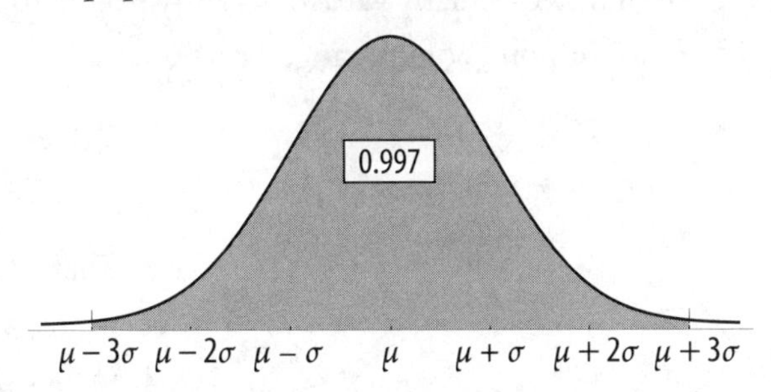


Remember, the total area under the whole curve from $-\infty$ to $+\infty$ is exactly 1, so you approach 100% but never get there.

Ultimately, what you always want to do is express the range you care about in terms of standard deviations from the mean. Imagine the following Socratic dialogue:

| | |
|---|---|
| Socrates: | "What percent of the population weighs between 144 and 155 pounds, if the weight is normally distributed?" |
| You: | "What's the mean? What's the standard deviation?" |
| Socrates: | "The mean μ is 150 pounds, and the standard deviation is 2 pounds." |
| You: | "Okay. The curve is centered on 150 pounds, and the yardstick is 2 pounds." |

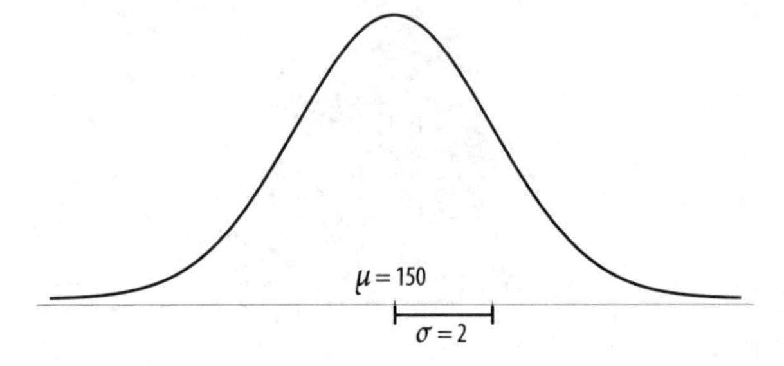


| | |
|---|---|
| You again: | "All right. How many yardsticks away from 150 are each of the endpoints of the range I care about?" |

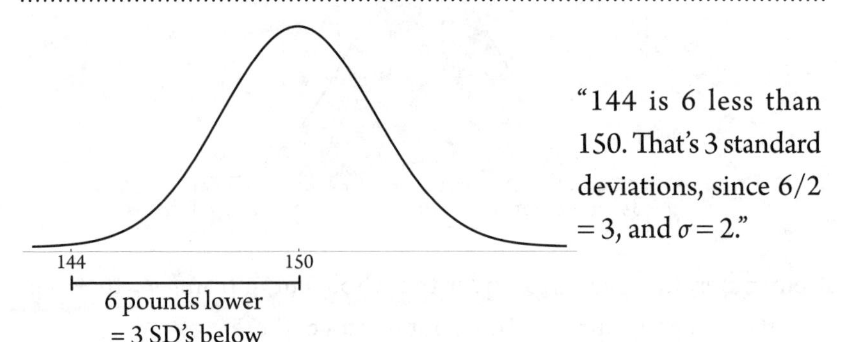


"144 is 6 less than 150. That's 3 standard deviations, since 6/2 = 3, and $\sigma = 2$."

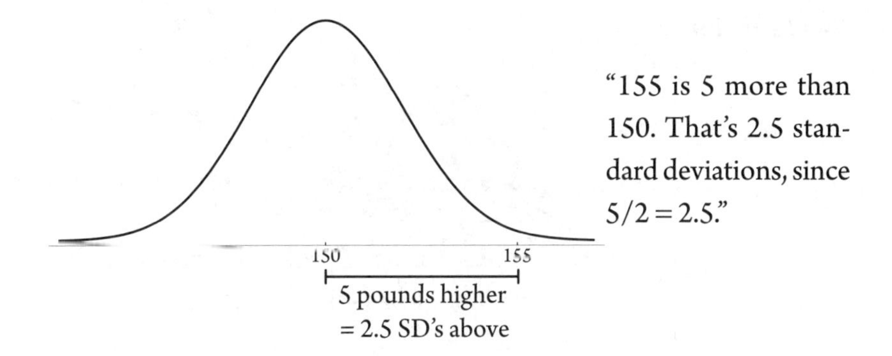


These counts of standard deviations are called ***Z-scores***.

Low end: 144 pounds = 3 SD’s below the mean → $Z = -3$

High end: 155 pounds = 2.5 SD’s above the mean → $Z = +2.5$

The Z-score of the mean $\mu$ itself is 0 (since it’s no distance away from itself), and the Z-score of $\mu + \sigma$ is 1 (that’s 1 SD above the mean). In a sense, what you’re doing is changing over to “Z-score land,” a happy place that is home to the ***standard normal curve***, where the Z mean is always 0 and the Z standard deviation is always 1. Computing a Z-score is called ***standardizing***.

**Original Curve**:

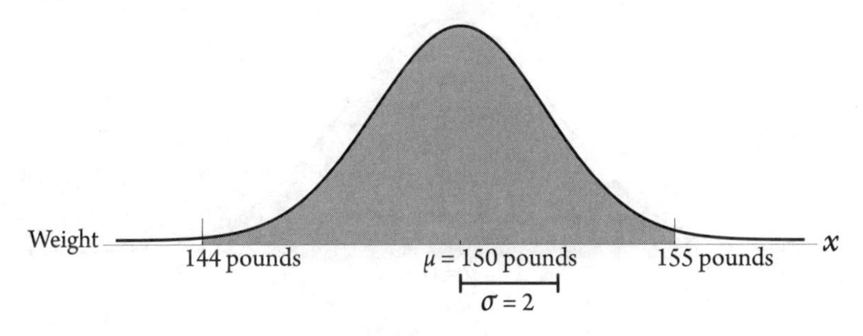

**Standard Normal:**

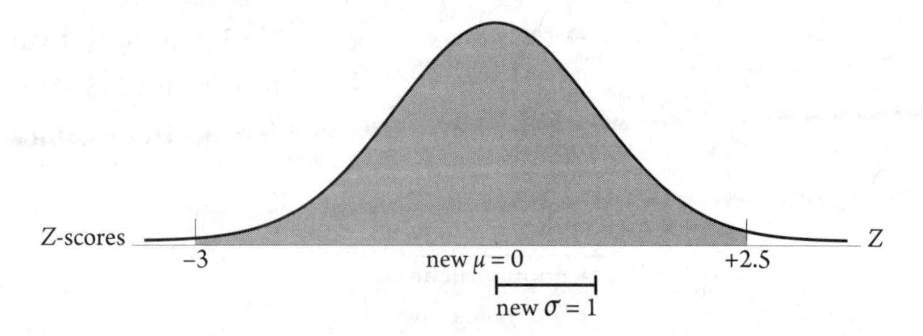


Remember, a $Z$-score just tells you how many standard deviations away from the mean some number is. Here's the formula:

| Endpoint | – | Mean | = | $Z$-score | . | SD |
|---|---|---|---|---|---|---|
| $x$ | – | $\mu$ | = | $Z$ | . | $\sigma$ |

Rearranging, you get this:

$$\frac{x-\mu}{\sigma}=Z$$

Now what? You have this:

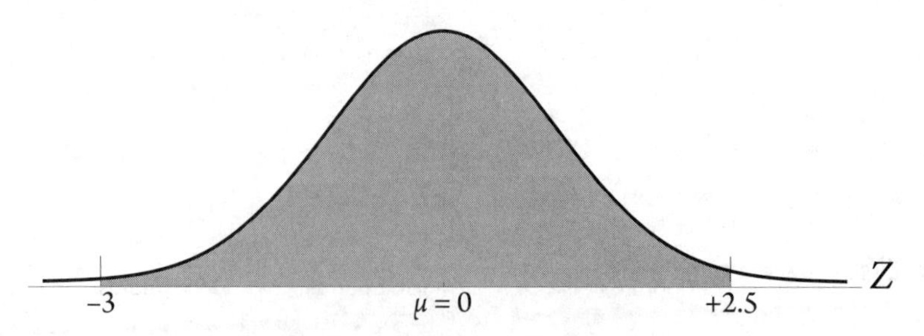


Now you look up the area in the back of your textbook, where you'll find a fancy-dancy table that lays out hundreds of $Z$-scores and areas. Or you can compute the area in Excel.

If you use a table, it's a two-step process because of the way the tables are always laid out—they only do half of the curve.

**Step 1: Right Half**

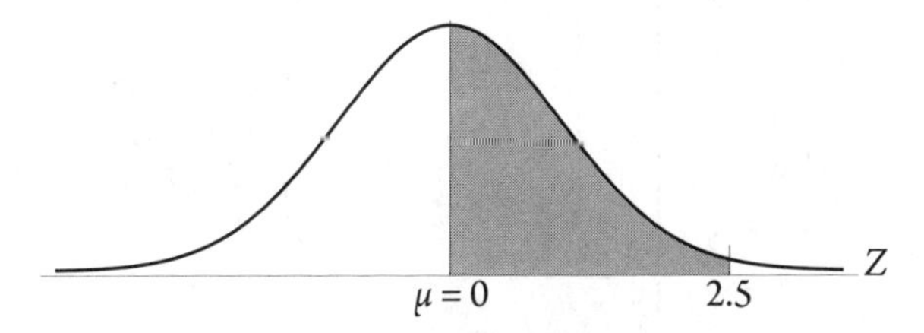


You want the area under the curve between $Z = 0$ and $Z = 2.5$. Many tables go to the hundredth's value, so 2.50 is the number you look up.

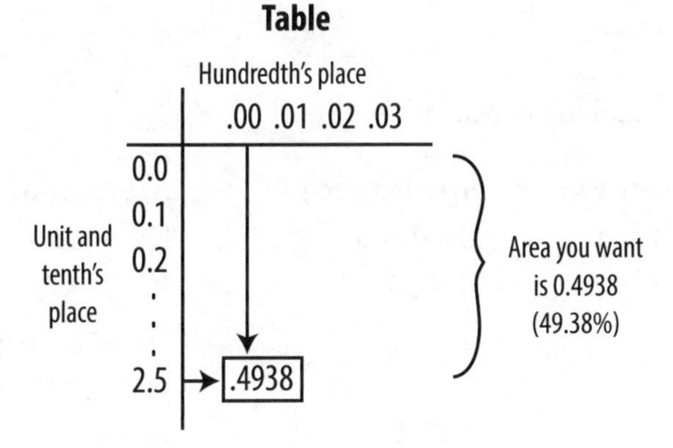


**Step 2: Left Half**

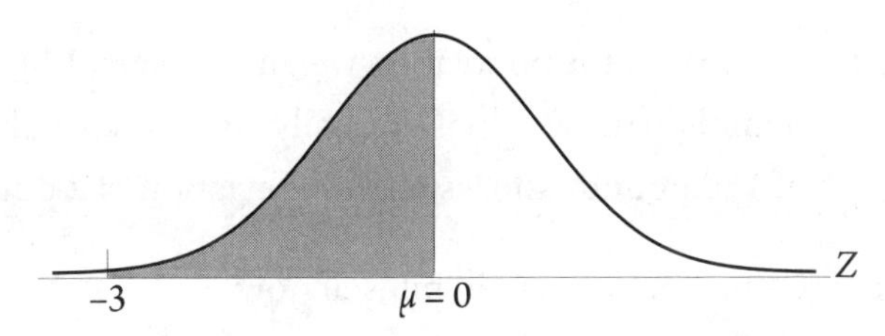


You want the area under the curve between $Z = -3$ and $Z = 0$. You don't need another table for negative values—the area is the same to the left and to the right of the mean. Again, look up the endpoint to the hundredth's place (3.00).

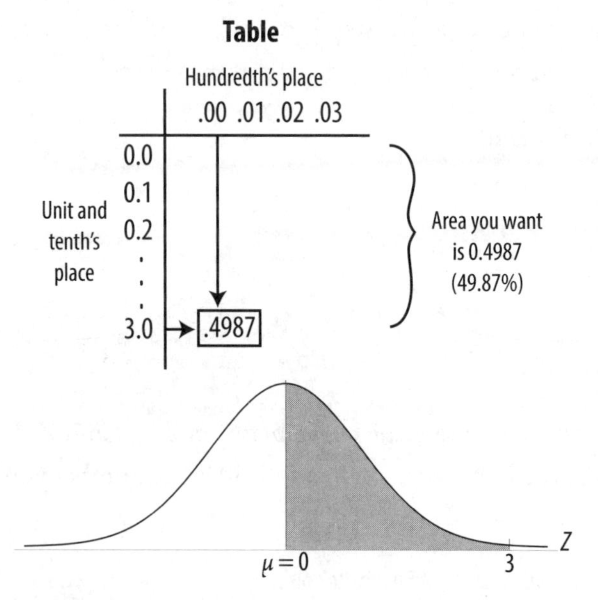


Finally, you add the two areas up.

| Area between $Z = -3$ and $Z = 2.5$ | = | Area between $Z = 0$ and $Z = 2.5$ | + | Area between $Z = -3$ and $Z = 0$ (same as area between 0 and 3) |
|---|---|---|---|---|
| | = | 0.4938 | + | 0.4987 |
| | = | 0.9925 (99.25%) | | |

| You: | "99.25% of the population weighs between 144 and 155 pounds, if the weight is normally distributed with a mean of 150 pounds and a standard deviation of 2 pounds." |
|---|---|

Whew! Socrates had better not kill himself now.

With Excel, it's still two steps, but the formulas work a little differently than the table does. The Excel function NORMSDIST gives you the ***cumulative*** area up to some Z-score, all the way from minus infinity ($-\infty$) on the far left. For example, NORMSDIST(1.2) computes the following area as 0.8849:

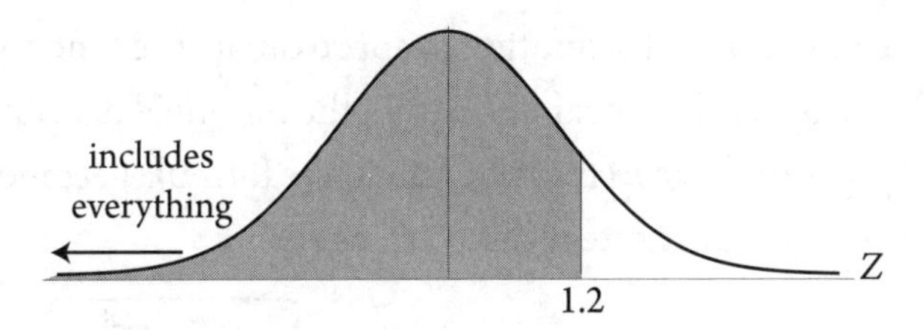


Say you want to find the area between $Z = -0.3$ and $Z = 1.2$. Compute the NORMSDIST value for the upper end of the range (1.2), then subtract the NORMSDIST value for the lower end of the range (−0.3). Don't flip the curve, as you would when working with a table. Keep the $Z$-scores negative.

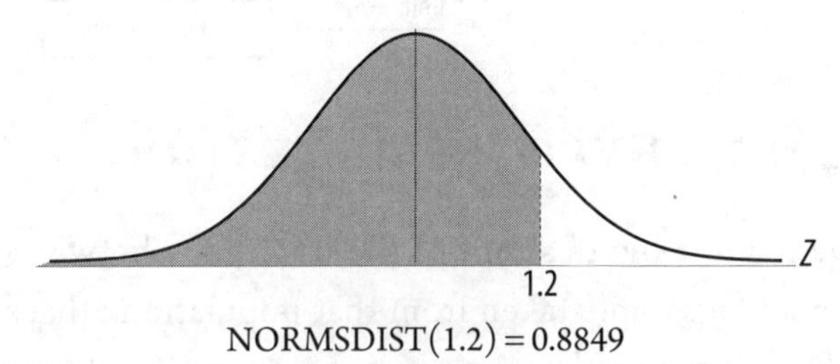


NORMSDIST(1.2) = 0.8849

**MINUS**

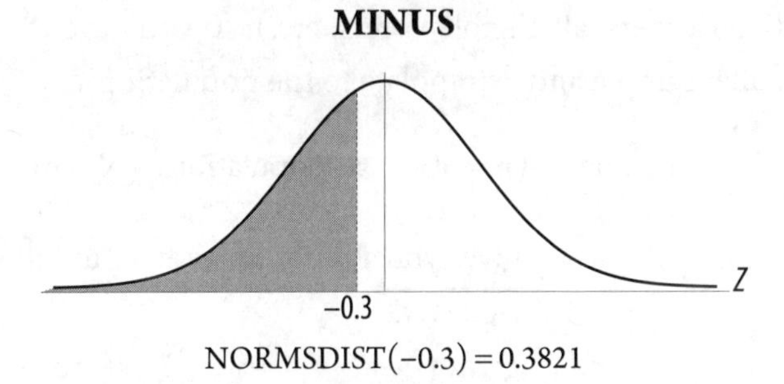


NORMSDIST(−0.3) = 0.3821

**EQUALS**

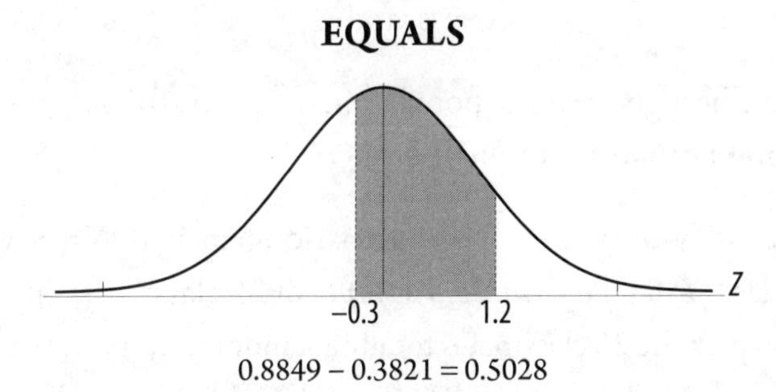


0.8849 − 0.3821 = 0.5028

You can use the normal distribution to approximate the binomial distribution if $n$ is large and $p$ is near 0.5. Again, the binomial distribution is a pain to calculate for large $n$ (the factorials in the formula get ginormous), so the normal can substitute when $p$ is close to ½.

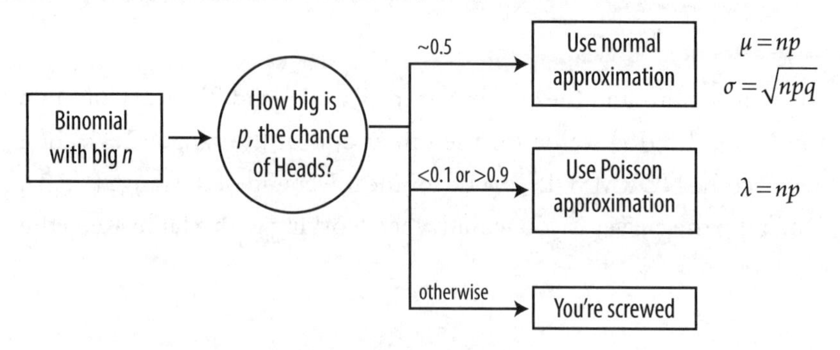


## SAMPLING & HYPOTHESIS TESTING

Earlier, we made a big deal out of the difference between a whole population and a sample taken from that population. That's because you usually can't survey the whole population you're interested in (all potential consumers, all eligible voters, etc.), so you have to make do with a smaller sample and extrapolate to the population.

| | Sample | ($n = 50$) | Population | ($N =$ millions) |
|---|---|---|---|---|
| Mean | $\overline{x}$ | gives you a → guess at | $\mu$ | (unknown) |
| Standard Deviation | $s$ | gives you a → guess at | $\sigma$ | (unknown) |

Your single best guess at the population mean $\mu$ is the sample mean $\overline{x}$, called a ***point estimator*** (a single best guess) for $\mu$.

Consider the "years since college" scenario again, with $N = 500$ people in your class. You take a ***random sample*** of 25 classmates ($n = 25$), in which you pick the 25 classmates totally at random. Everyone has to have an equal chance of being picked. In the real world, truly random samples are difficult to generate, but they're crucial for accurate results—if the sample isn't random, all bets are off. This is one reason why pharmaceutical drug trials are extremely costly to run.

If the true (but unknown) population mean $\mu$ is 3.0 years, then the expected value of the sample mean will be—surprise!—3.0 years as well. That is, you'd properly expect your sample mean to be 3.0 years.

| $E(\overline{x})$ | $=$ | $\mu$ |
|---|---|---|
| Expected value of sample mean | | Population mean |

Imagine you did 100 different experiments and took 100 different random samples, each one with 25 people. Measure the sample mean $\overline{x}$ for each of these samples.

| | | |
|---|---|---|
| Sample 1 | $\overline{x} = 3.2$ | Average should be 3.0, the real mean of the population. |
| Sample 2 | $\overline{x} = 2.7$ | |
| Sample 3 | $\overline{x} = 2.9$ | |
| ... | $\overline{x}$... | |
| Sample 100 | $\overline{x} = 3.4$ | $\mu_{\overline{x}} = \mu_{\text{pop}}$ |

How spread out would those 100 sample means be? That is, what would their standard deviation be? The answer depends on two factors.

1. The *standard deviation of the population* you're taking the samples from. This is the raw source of any spread in the sample means.

If the population contains widely varying data, then the samples extracted from that population will have widely varying means.

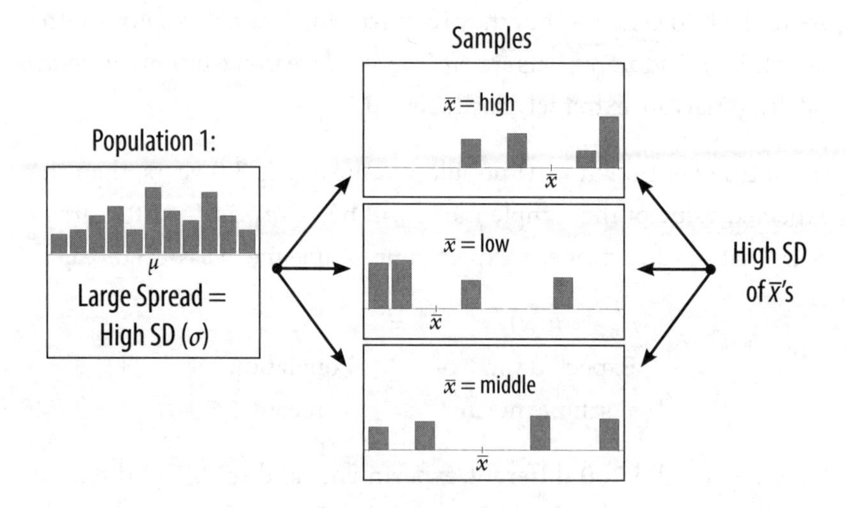


On the other hand, if the population numbers are all packed in tightly, then you simply can't generate samples very far apart from each other, so the sample means will have a low standard deviation.

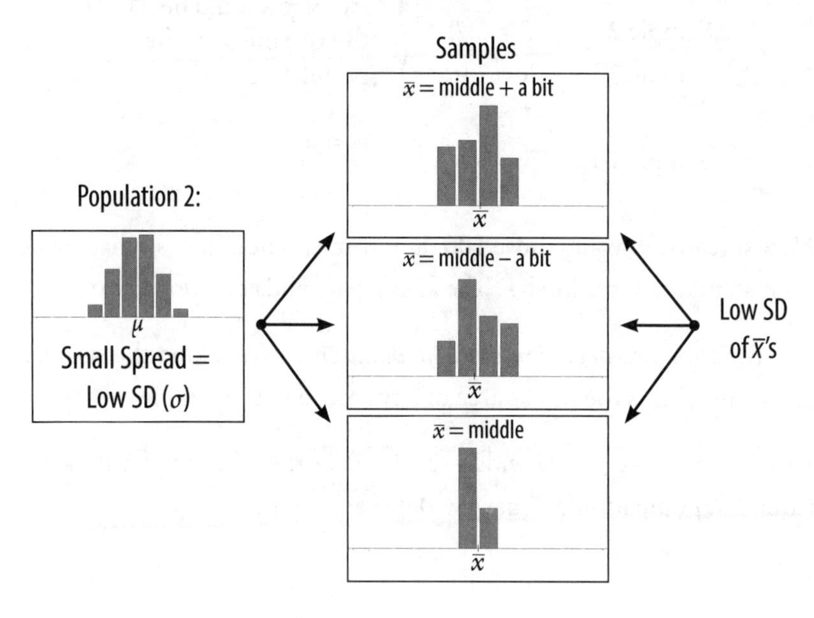

2. *The value of* n *for each sample*—how big each sample is. The larger $n$ is, the tighter the $\bar{x}$'s will bunch around the population mean. That's because the more data you have *in each sample*, the more likely you are to draw from all parts of the population. Each additional data point makes a sample more representative of the population.

If you take small samples, say with $n = 7$, then there's a good chance that within any given group of 7 points, you pick a bunch from one end or the other of the population, causing the mean to be skewed. Thus, the sample means will be highly variable.

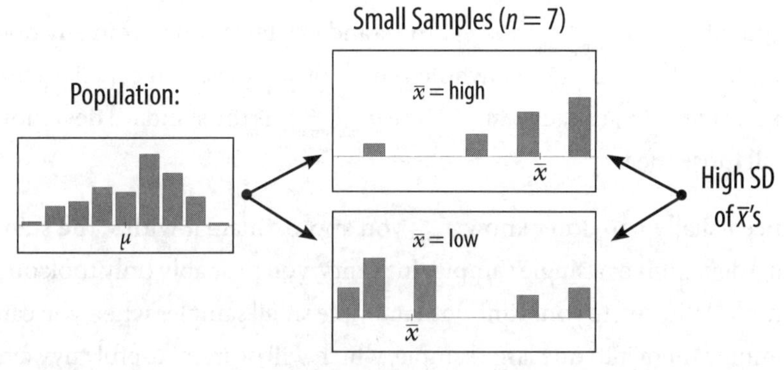


In contrast, if you take large samples, say with $n = 50$, then each sample is more likely to represent the whole population and therefore to have a mean close to the true population mean. The sample means will be bunched tightly together.

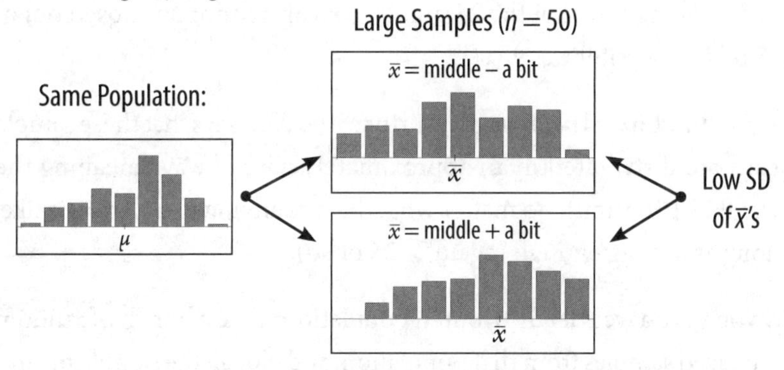

The standard deviation of the sample means is determined by the SD of the population and the size of the sample. The precise mathematical relationship is:

$$\sigma_{\text{sample means}} = \frac{\sigma_{\text{population}}}{\sqrt{n}}$$

$$\sigma_{\overline{x}} = \frac{\sigma_{\text{pop}}}{\sqrt{n}}$$

Remember that $\sigma_{\text{sample means}}$ is *not* the standard deviation within any one sample. Rather, you grab a whole bunch of samples. You calculate the sample mean ($x$) of each sample. The $\sigma_{\text{sample means}}$ is the standard deviation of all those means.

Since usually you don't know $\sigma_{\text{pop}}$, you approximate it with *s*, the standard deviation of a single sample. In reality, you probably only took one sample. Why waste your time with multiple small samples when you can combine them into one large sample, which will be more useful anyway? Approximating $\sigma_{\text{pop}}$ with *s* is fine, so long as *n* is large enough (> 25 or so). The definition of "large enough," vague as it may seem, doesn't depend on the size of the original population. A sample of 25 data points works for a population of 500 or of 5 billion. (What makes the second case hard is making sure that the 25 points are truly randomly chosen out of all 5 billion possibilities.)

The ***Central Limit Theorem*** goes further. The ***CLT*** says that these sample means are distributed in an approximately *normal* way (meaning the normal distribution), no matter what the underlying population is like, as long as *n* is big enough (again, > 25 or so).

Say you have a weirdly distributed population. Take a bunch of random good-sized samples from that population, and plot all the sample means. Magically, that plot will look like a normal distribution. This is why it's so "normal."

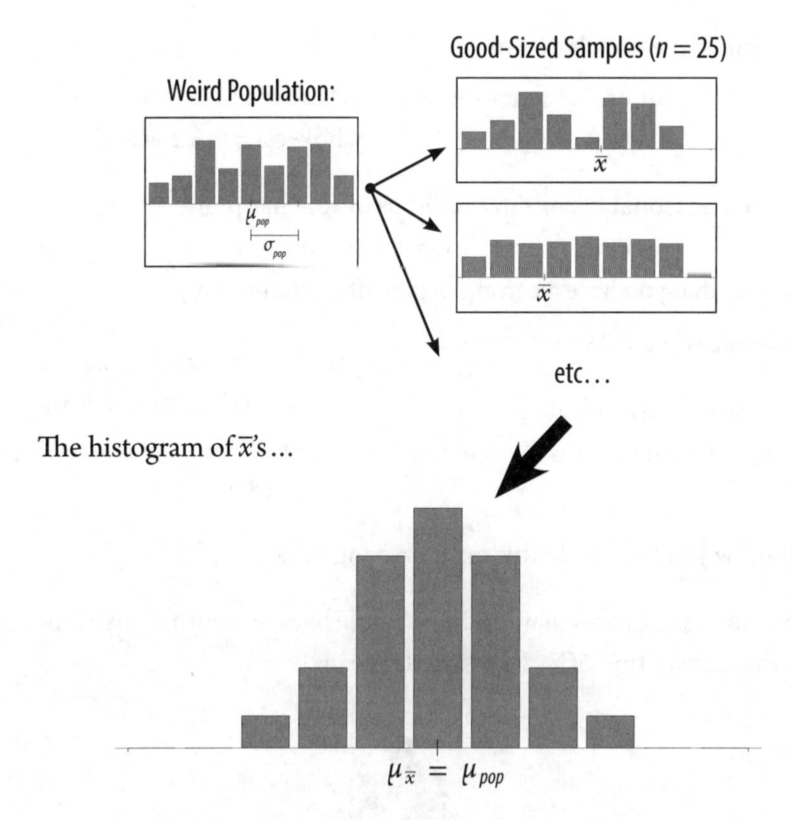


The histogram of $\bar{x}$'s...

...fits a normal distribution.

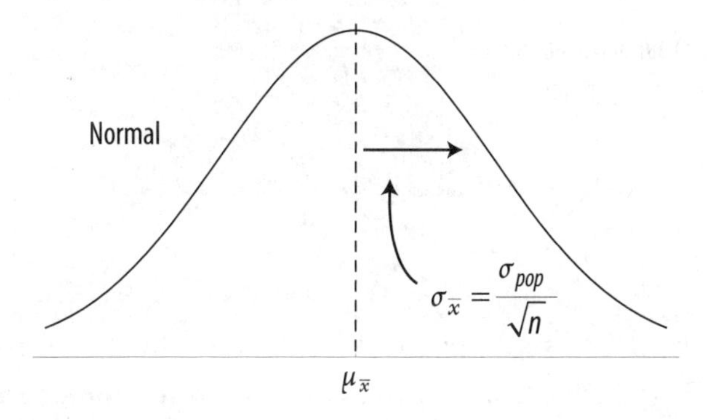


Why does all this matter?

Because it lets you take just one good-sized random sample and do two things:

1. Estimate the population mean.

   Sample mean $\overline{x} = 3.2$ years ⟶ You: "I think the mean $\mu$ for the whole class is 3.2 years."

2. Give a reasonable ***confidence interval*** for that point estimate. A confidence interval is exactly what it sounds like—it is a range of values that you have a certain degree of confidence in.

   Sample size $n = 25$

   Sample standard deviation $s = 0.1$ (computed on the sample)

   *magic* ⟶ You: "I'm 95% sure that the mean for the whole class is between 2.8 years and 3.6 years."

We'll show you how to do this magic—hang tight.

The mean of your particular random sample lives in a normal distribution: the distribution of $\overline{x}$'s (the sample means).

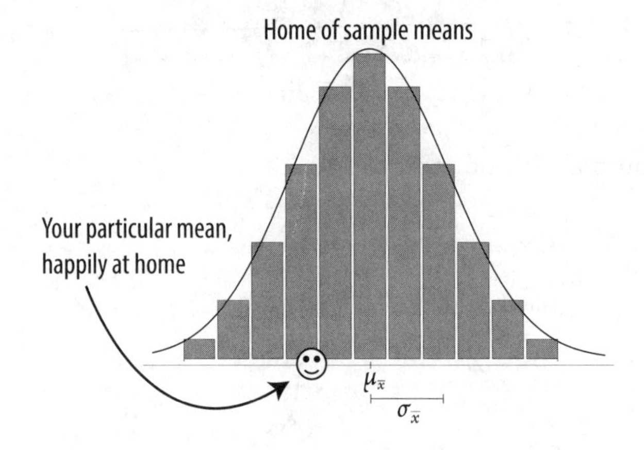


The mean of all the $\overline{x}$'s is the population mean, and the standard deviation of all the $\overline{x}$'s (which is also known as the ***standard error***, or ***SE***, so that you don't confuse it with the population standard deviation) is given by that formula from earlier:

$$\sigma_{\overline{x}} = \text{SE} = \frac{\sigma_{\text{pop}}}{\sqrt{n}}$$

All the nice properties of the normal distribution can be applied here.

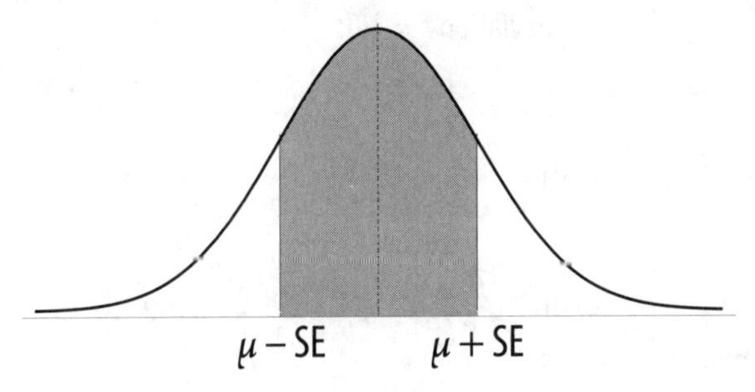


~68% of sample means will be within 1 standard error of the population mean $(-1 \le Z \le 1)$.

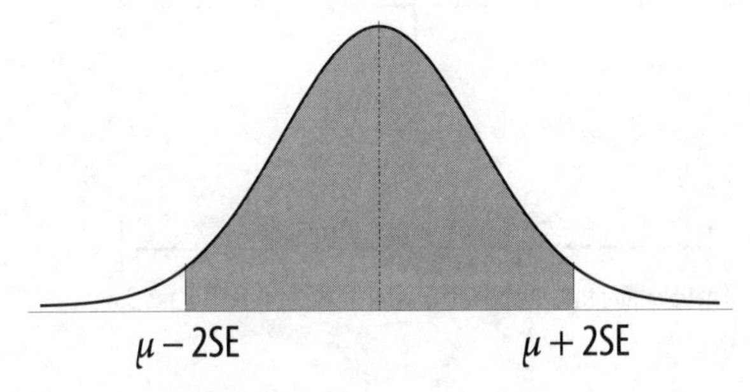


~95% of sample means will be within 2 standard errors of the population mean $(-2 \le Z \le 2)$.

You can flip around this thinking and say this:

- If you just know a single sample mean, 95% of the time, the population mean will be within 2 standard errors of it.

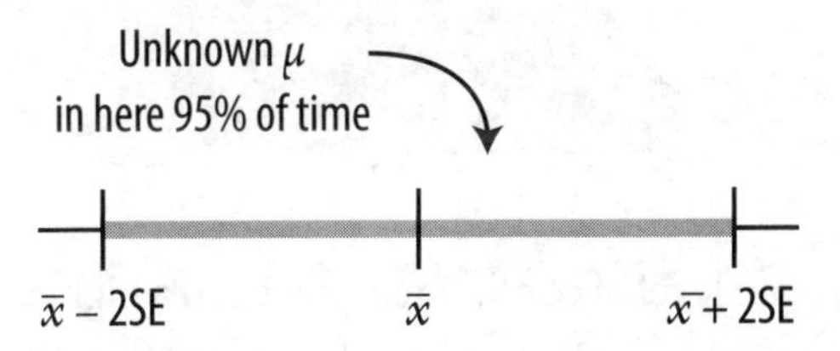


All you have to do is compute the standard error, then we can set up the confidence interval.

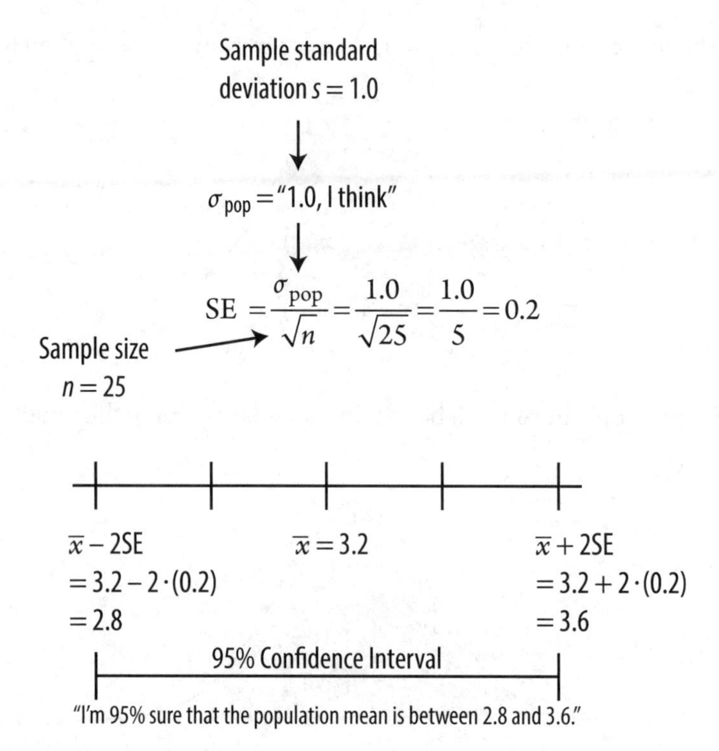


By the way, 1.96 is often used instead of 2 for the $Z$-score corresponding to a 95%-confidence level, since 1.96 is a little more accurate.

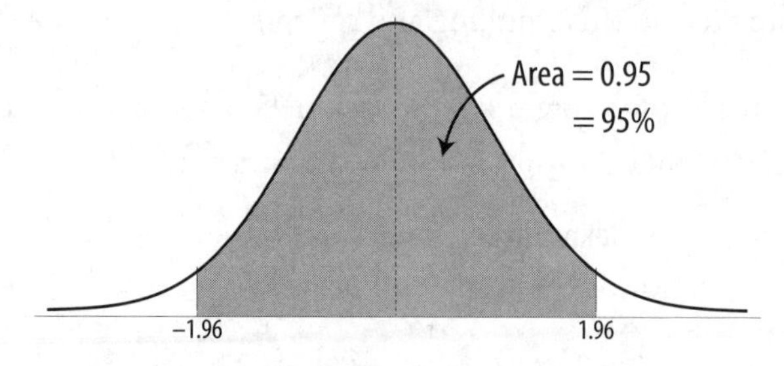


If you want a higher level of confidence, you need a higher $Z$-score (more standard errors away from the mean), which means a wider interval. For instance, a range of ±3 for $Z$ encloses 99.7% of the area under a normal curve, so ±3 SE's gives you a 99.7% confidence interval.

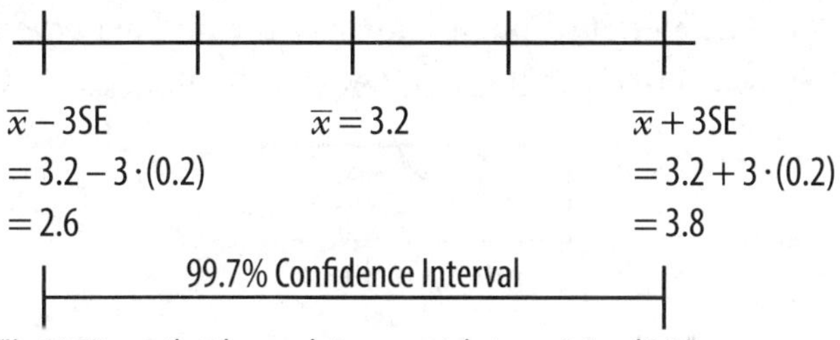


## SMALL SAMPLE

If $n$ is small and you don't know the population $\sigma$ (which you almost never do), you're up a creek—unless you know or have reason to assume that the underlying population is normally distributed. If so, the sample means for small samples follow something called a ***t distribution***, which is a lot like the normal distribution, except it's got bigger, fatter tails. The $t$ distribution you use depends on the exact sample size too—the smaller $n$ is, the flatter and more spread out $t$ is.

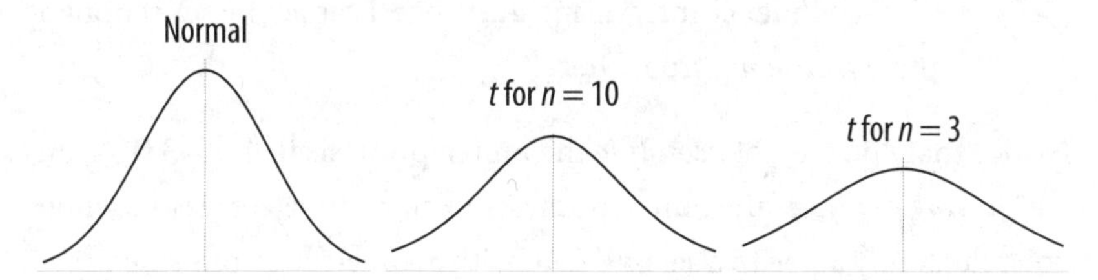


You look up and use $t$ scores the same way you use $Z$-scores—to find a confidence interval for your estimate of the population mean.

## HYPOTHESIS TESTING

The 95% confidence interval in the earlier example was 2.8–3.6, meaning that you're 95% sure that the real population mean is somewhere between 2.8 and 3.6.

If you had the prior hypothesis that the population mean was 2.2, then you could ***reject*** that hypothesis with 95% certainty. The prior hypothesis, a kind of default position, is known as the ***null hypothesis*** or $\boldsymbol{H_0}$. You typically want to reject this uninteresting hypothesis.

$$\boxed{H_0 : \mu = 2.2}$$

null hypothesis

"I reject thee with 95% certainty, because my 95%-confidence interval does not include thee. I am 95% sure that $\mu$ is between 2.8 and 3.6."

You might not be able to reject a different null hypothesis, though.

$$\boxed{H_0 : \mu = 3.0}$$

another null hypothesis

"I would like to reject thee, since my $\overline{x} = 3.2$; however, since my 95%-confidence interval includes thee, I cannot be sure enough that you are *not* true. Drat!"

Notice that you are not *accepting* the null hypothesis that $\mu = 3.0$—you are "***failing to reject***" the null hypothesis. "Failing to reject" is like when you reluctantly agree to a second date with a none-too-promising prospect, but you go along with it because you have nothing better to do.

There is also an ***alternative hypothesis***, $\boldsymbol{H_a}$, which is what you wish to support instead of the null. Say $H_0$ proposes that $\mu = 3.0$. Then $H_a$ could simply be that $\mu$ *doesn't equal* 3.0. If that's the case, you need a ***two-tailed test***, which allows you to reject the null if the result is *either* a lot larger than 3.0 *or* a lot smaller than 3.0.

In contrast, you have a ***one-tailed test*** if your $H_a$ proposes only that $\mu$ *is greater than* 3.0. For some reason, you won't be able to reject the null if $\mu$ winds up being extremely low. Similarly, you have a one-tailed test if your $H_a$ proposes specifically that $\mu$ *is less than* 3.0.

This issue may seem tricky, but which kind of test to use should be apparent in the wording of the problem. Just think about whether you'd like to "prove" that the mean is different from a specified number (two-tailed test) or only that it falls to a particular side of that number (one-tailed test).

| $H_0 : \mu = 3.0$ | Type of Test | Tails |
|---|---|---|
| $H_a : \mu \neq 3.0$ | Two-tailed | |
| $H_a : \mu > 3.0$ | One-tailed | |
| $H_a : \mu < 3.0$ | One-tailed | |

To test a hypothesis:

1. Define $H_0$ and $H_a$ first. Usually the problem specifies an assumed mean, $\mu_0$, and also indicates whether you need a two-tailed or a one-tailed test.

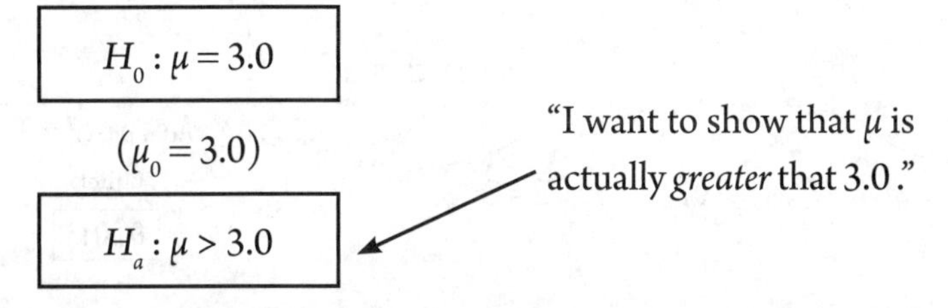


2. Take your random sample of size $n$ and calculate $\overline{x}$ and $s$. Say you get these results:

$$\overline{x} = 3.3 \qquad s = 1.0 \qquad n = 100$$

3. Figure out how many standard errors (SE) the $\overline{x}$ is away from your assumed null mean $\mu_0$. This tells you how "close" or "far" $\overline{x}$ is.

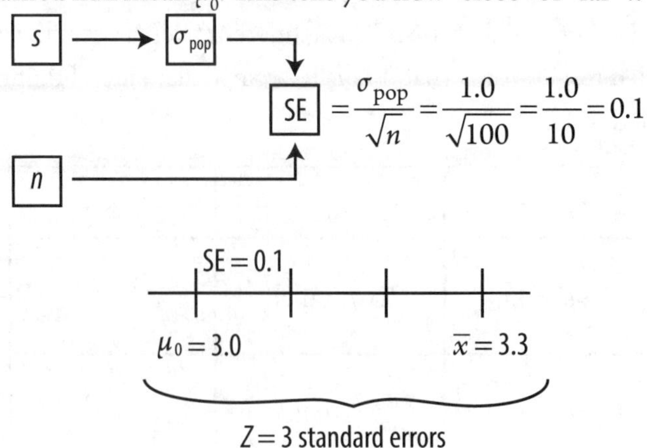


The one-step formula is:

$$Z = \frac{\overline{x} - \mu_0}{SE} = \frac{\overline{x} - \mu_0}{\sigma_{pop} / \sqrt{n}} = \frac{\overline{x} - \mu_0}{s / \sqrt{n}}$$

4. Now, for a one-tailed test, look up how much area would lie farther away from 0 on a standard normal distribution than that $Z$-score.

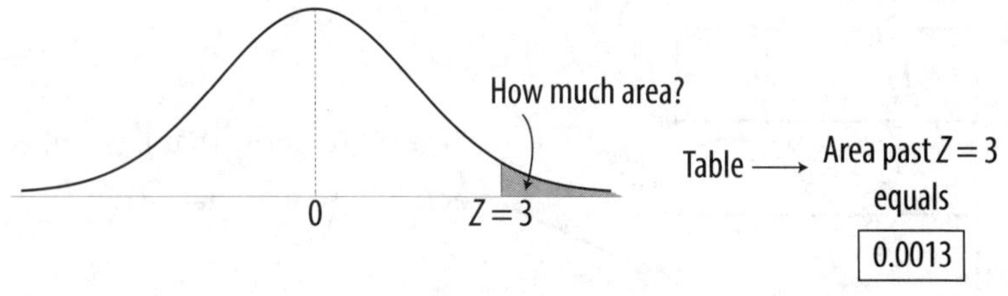


That 0.0013 is known as the **p-*value*** of this test. (No connection to the little $p$ from earlier, with the binomial.) If $p = 0.0013$ for this test, it means that there's only a 0.0013 chance (0.13%) that the following situation has occurred: the null hypothesis is true (i.e., $\mu$ is really 3.0) *and* you get $\overline{x} = 3.3$ with your sample, together with the other stats. 3.3 might not seem all that much bigger than 3.0, but since the standard error is so small (only 0.1), that 3.3 is a full 3 standard errors above the null mean.

That's very unlikely ***unless*** the null hypothesis is wrong. So now you can take the final step, figuring out where you stand on the null hypothesis.

5. Decide on a confidence level, and either reject $H_0$ or "fail to reject" $H_0$.
   - Wanna be 90% sure? Okay—reject.

     90% sure = 10% chance of error is okay for you. This acceptable error level is $\boldsymbol{\alpha}$, the Greek letter alpha.

     Since $p < \alpha$, you can safely reject it. $0.0013 < 0.10$.
   - Wanna be 99% sure? Okay—reject.

     $\alpha =$ only 0.01, but $p$ is still smaller.
   - Wanna be 99.9% sure? Hmm—sorry! You can't be that sure with the results of this experiment. $Z$ is large (good), so $p$ is small (good), but not small enough for your finicky tastes. If you have a very small $\alpha$ (here, $\alpha = 0.001$), then you'll need an even smaller $p$-value in order to reject the null hypothesis.

Long story short, there are two ways a hypothesis test can shake out, given a particular confidence level.

1. Good enough to reject $H_0$ ($\bar{x}$ is "far" away from $\mu_0$ as measured by the SE yardstick).

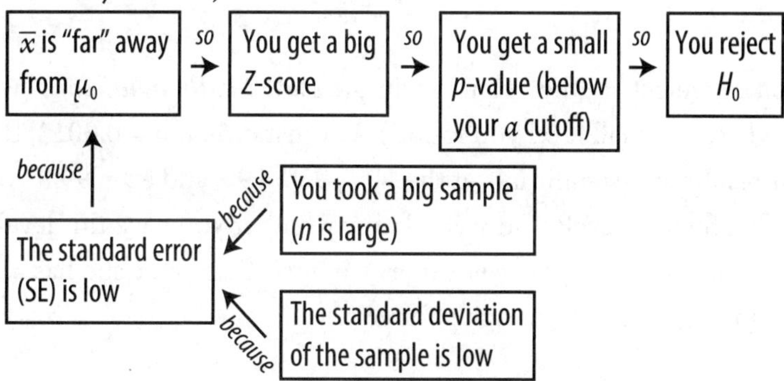

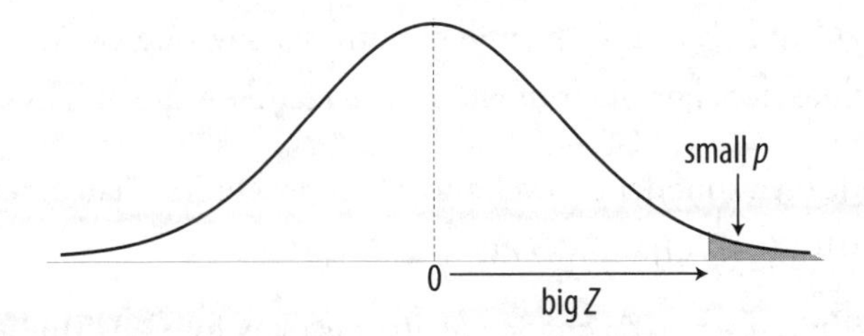


2. Not good enough to reject $H_0$ ($\bar{x}$ is "close" to $\mu_0$ as measured by the SE yardstick).

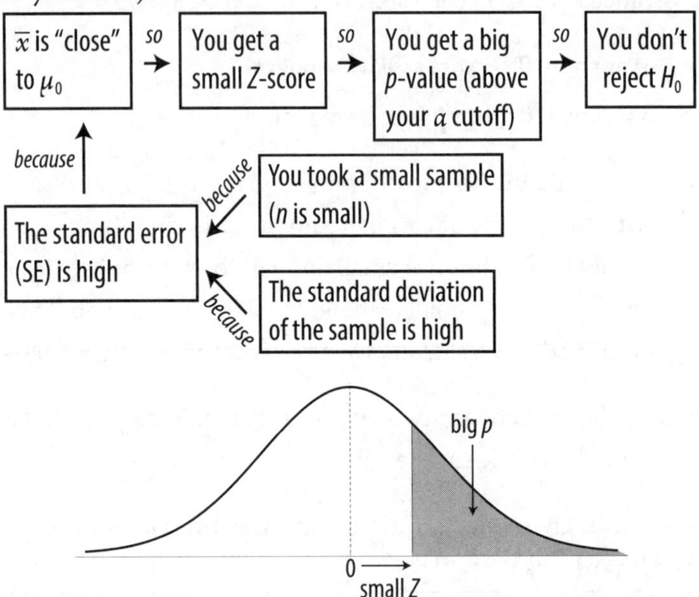


If you can reject $H_0$, then your results are ***statistically significant*** (with $p$ = whatever small level you found). For instance, if $p = 0.0013$, then your results are "significant" at the 90th , 95th, 99th, and even 99.8th percent confidence level. (Only you know whether you have this level of confidence in your *significant* other; we hope that he or she has a big enough $Z$ and a small enough $p$ for you.)

## CORRELATION AND REGRESSION

Up to now, everything has had to do with one variable—one measurement, one kind of data.

| Student | Years since College |
|---|---|
| You | 4 |
| Alice Atwater | 2 |
| Bill Burns | 6 |
| ... | ... |

One variable for each observation

What if you get *two* pieces of data about each person? Now you can look at patterns.

*Example 1: Years since College and Height*

| Student | Years since College | Height |
|---|---|---|
| You | 4 | 5 feet 7 inches |
| Alice Atwater | 2 | 5 feet 10 inches |
| Bill Burns | 6 | 5 feet 2 inches |
| ... | ... | ... |

To look for a pattern, you put all these observations on a scatterplot:

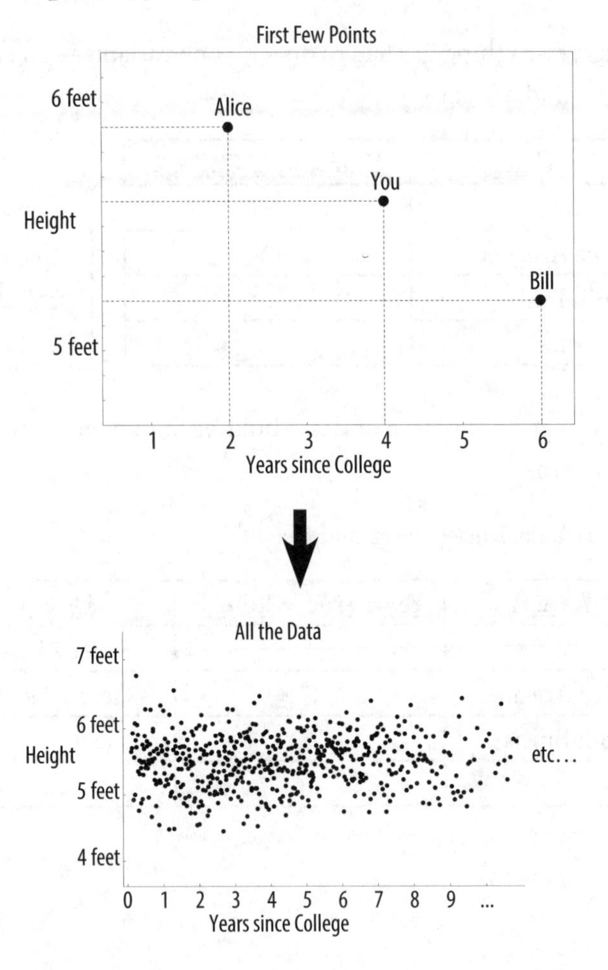


The second graph would almost certainly reveal no overall pattern. The "shotgun blast" shows that the two variables, Years since College and Height, are basically ***uncorrelated***.

*Example 2: Years since College and Age*

In contrast, if you plot Years since College versus Age, you'll get a pattern.

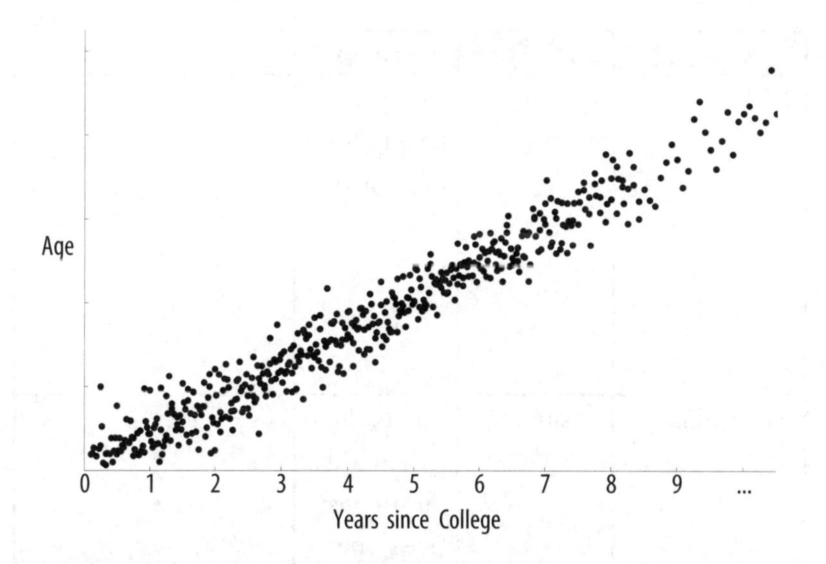


Most people are either high on both scales or low on both scales. You don't run into many Doogie Howsers (young age but many years since college) or Rodney Dangerfields[3] (old age but few years since college). This means that Years since College and Age are highly ***correlated***.

You can measure the degree of correlation with ***r***, the ***correlation coefficient***. Don't worry about how to crunch *r*—Excel will do it for you with the CORREL function. The important thing is to understand what *r* tells you.

3 As Thornton Melon in the movie *Back to School*.

| Range of $r$ | Correlation | Pattern | |
|---|---|---|---|
| 1 (max) | Perfect positive | All points lie on a line of positive slope | 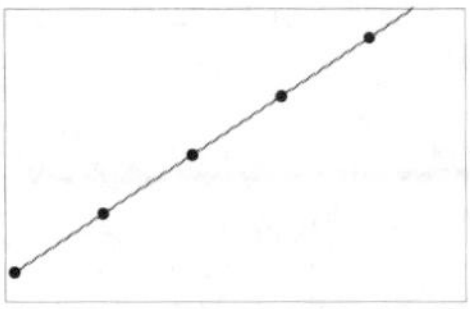 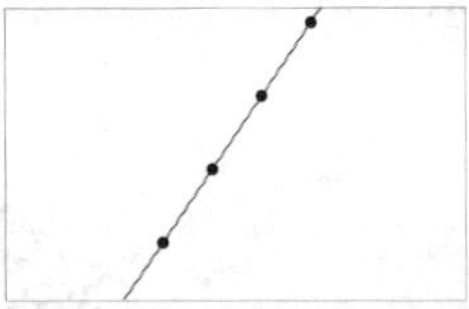 |
| 0.*something* | Positive | Points cluster around a line of positive slope | 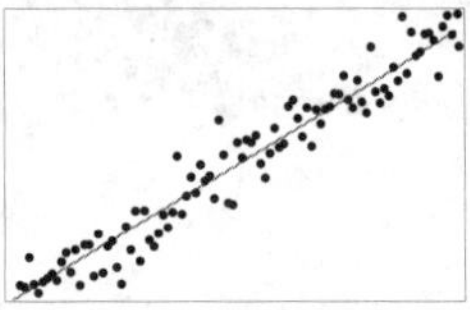 |
| 0 | None | No pattern (shotgun blast) *or* a nonlinear pattern | 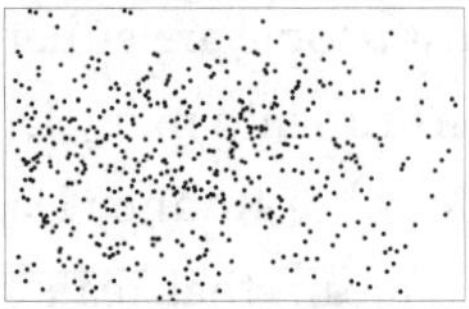 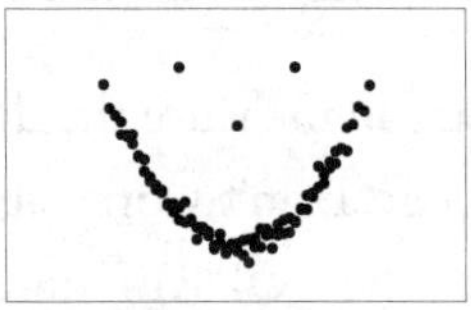 |
| −0.*something* | Negative | Points cluster around a line of negative slope | 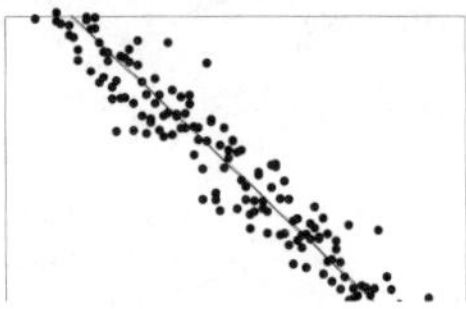 |
| −1 (min) | Perfect negative | All points lie on a line of negative slope | 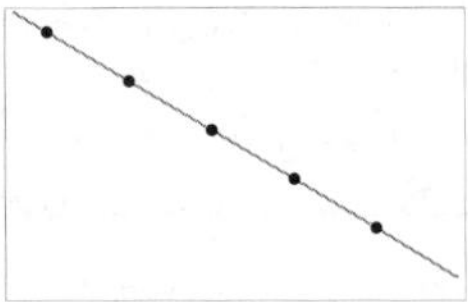 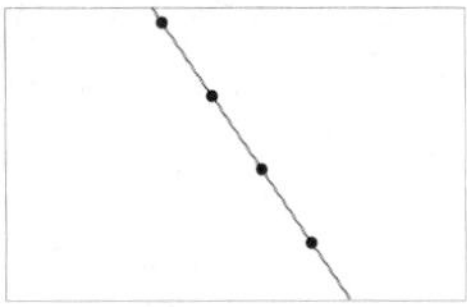 |

Notice that $r$ doesn't tell you the *slope* of the line. What $r$ tells you is the *sign* of that slope and more importantly *how tightly the points cluster* around a line of positive or negative slope.

Here's another way to think about correlation. Say you pick John Johnson out of the population, and John Johnson has a relatively *low* number of Years since College (below the mean). What would you expect for the levels of his other stats—Low, High, or No Idea?

**Age**: Positive Correlation ($r > 0$)

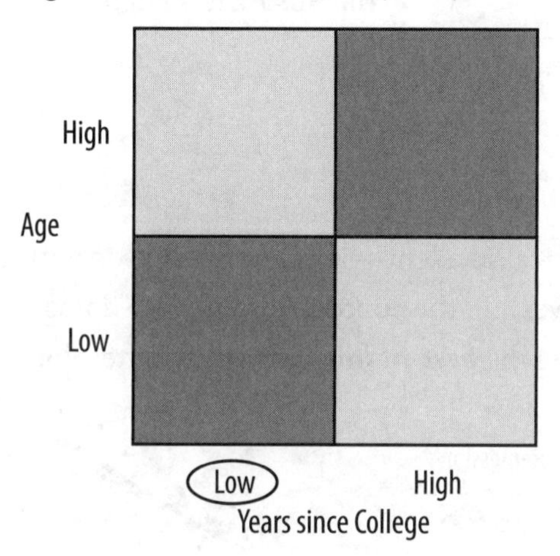


If you plot everyone on a two-by-two grid, you see large concentrations in the darker areas. So you expect John to have a Low age (that is, he's young). The higher $r$ is, the higher the probability that he's Low on Age.

**Height**: Zero Correlation ($r = 0$)

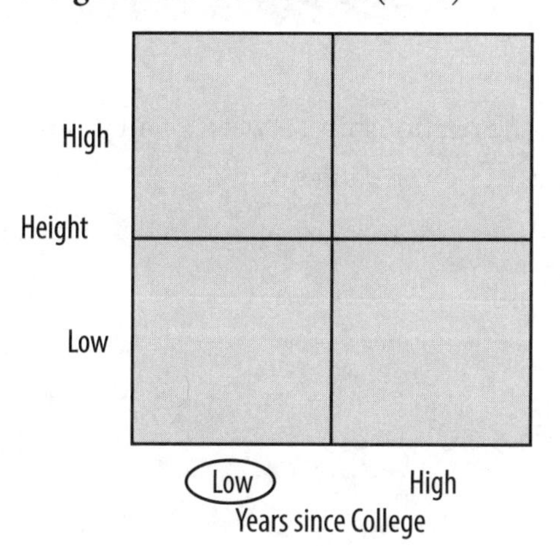


On this grid, there's no pattern. You have No Idea what to predict for John's height. Knowing his Years since College tells you zilch about his height.

**Familiarity with Facebook** (FwF): Negative Correlation ($r < 0$)

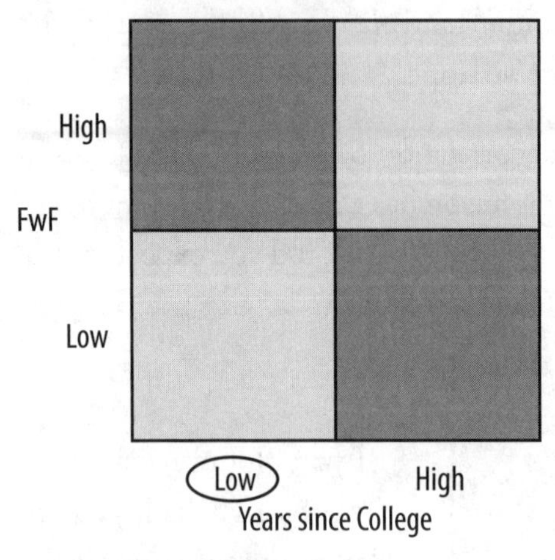


This plot shows lots of High-Low matchups. Since John is Low on Years, you expect him to be High on Facebook (so to speak). The closer $r$ is to $-1$, the higher the probability is that his FwF is High.

## REGRESSION

The name ***regression*** is a historical accident—it has nothing to do with reliving high school (or past lives). In the context of statistics, "doing a linear regression" means finding the ***best-fit line*** through a scatterplot.

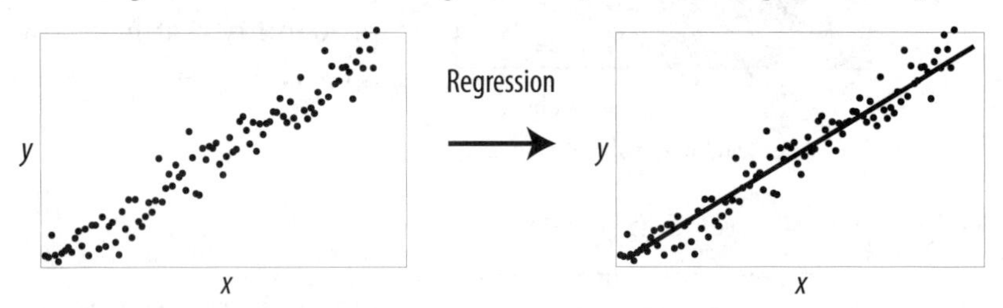


You do this so you can describe the relationship between $x$ and $y$ more precisely and even predict values of $y$ given values of $x$.

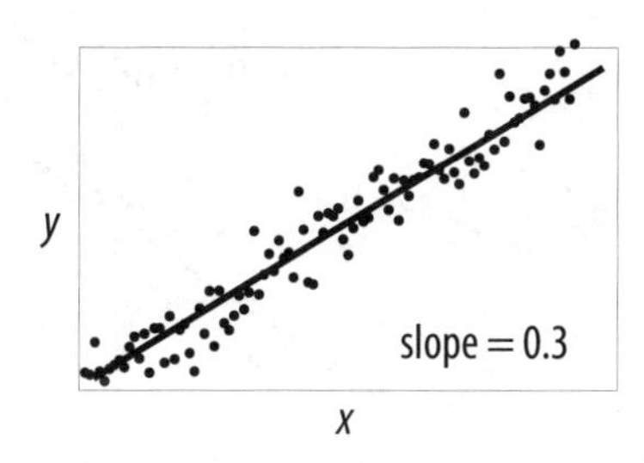


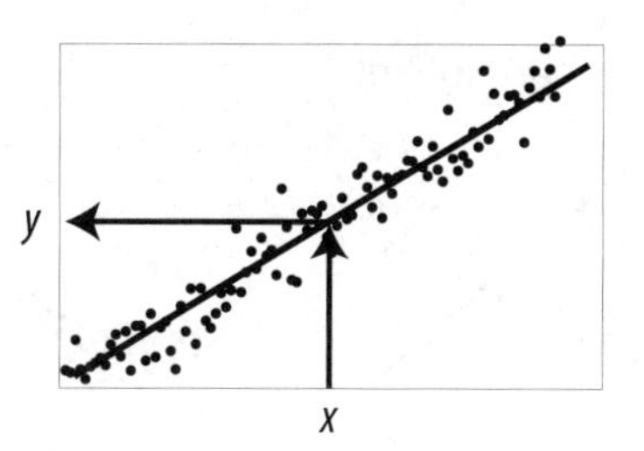


"If $x$ increases by 1, $y$ goes up by 0.3 on average."

"If $x = 30$, I predict that $y = 50$."

A few questions pop up right away:

- *Which variable should be x, and which should be y?*

  On a plain old scatterplot, it doesn't matter much, but when you regress, you want to make $x$ the ***independent*** variable (the input or the ***predictor*** variable). Then $y$ is the ***dependent*** variable (the output or the ***predicted*** variable). This could reflect that $x$ somehow causes $y$, or it could simply mean that $x$ is the easier thing to measure or control in the real world, while $y$ represents the ultimate phenomenon that you care about and want to explain in terms of $x$. For instance, if you have Sales and Price as variables, you'd make Price your $x$ variable and Sales your $y$ variable. Sales of a product are a function of the Price of that product, and you want to explain Sales in terms of Price.

- *What does "best fit" mean?*

  The best-fit line minimizes the overall distance, in some sense, between the points and the line.

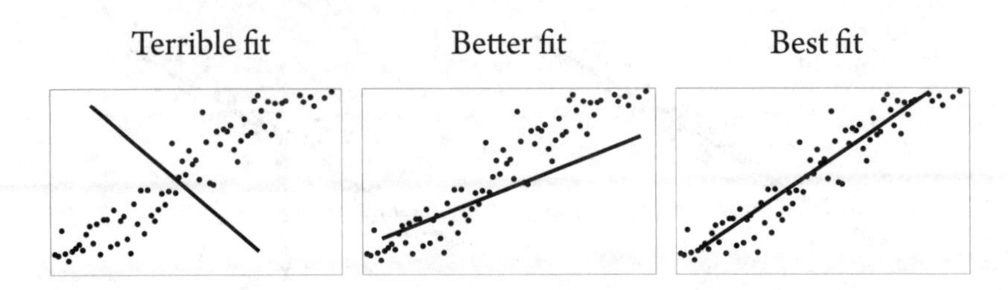


In standard linear regression, you measure the overall distance between the points and the line this way:

1) Measure each deviation or error: the vertical distance to the line

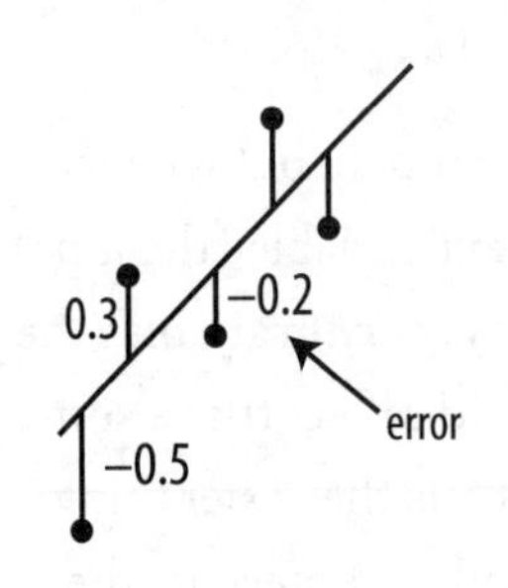


2) Square each of those errors and add them all up

$$(-0.5)^2 = 0.25$$

$$(0.3)^2 = 0.09$$

$$(-0.2)^2 = 0.04$$

$$\dots$$

$$= \boxed{9.83}$$

***Sum of squared errors*** is the "distance."

The best-fit line has the smallest possible sum of squared errors. This is why this line is often referred to as the ***best squares*** line.

- *What does it mean to "find" a line?*

  The general equation of a line is $y = mx + b$, as you recall from the GMAT or GRE. The letter $m$ represents the ***slope*** of the line, while $b$ represents the **y-*intercept***, where the line crosses the $y$-axis.

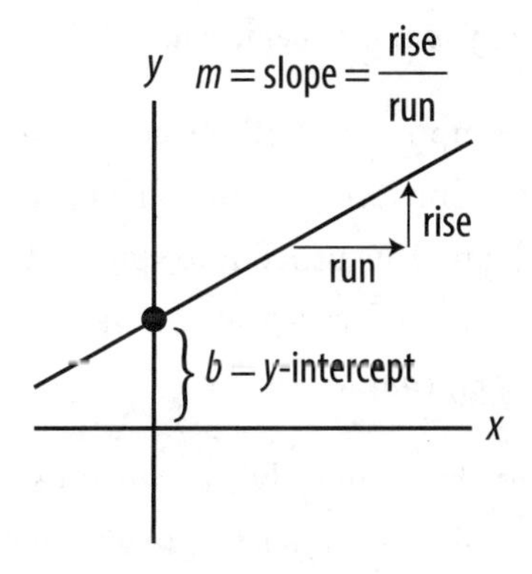


A line with equation $y = 3x - 2$ intercepts the $y$-axis at $(0, -2)$, and if $x$ increases by 1, $y$ increases by 3:

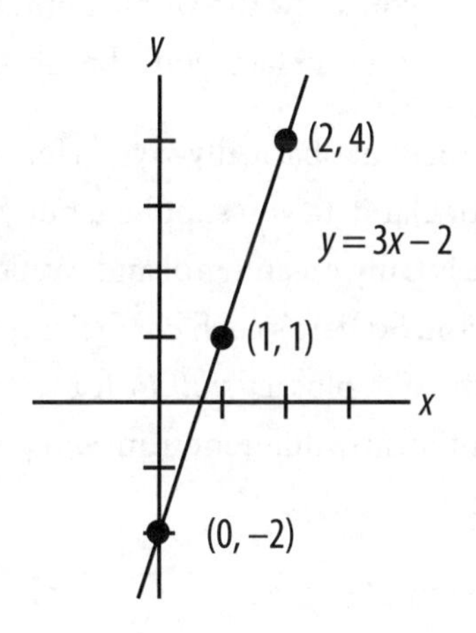


Finding the line means finding the values for the slope $m$ and the $y$-intercept $b$. In "Statistics Land," different letters are often used, unfortunately:

Instead of $y = mx + b$

you have $y = \beta_1 x + \beta_0$ or $y = \beta_0 + \beta_1 x$

$\boldsymbol{\beta}$ is the Greek letter ***beta*** (also used in finance, with a related but distinct meaning). So $\boldsymbol{\beta_0}$ (beta-zero) is the $y$-intercept, while $\boldsymbol{\beta_1}$ (beta-one) is the slope. These are collectively known as the ***coefficients*** of the regression line.

- *How do you actually find the coefficients?*

  Not to worry—computers do all the computations. In Excel, you call up a Regression procedure from a Data Analysis "toolpak" (an add-on package), and this procedure spits out not only the coefficients but also other information you need to tell whether those coefficients are *significant.*

  After all, the procedure can only *estimate* those coefficients from the data you give it—it assumes you're giving it a *sample,* not the whole population. (This should make sense—the "whole population" would include all future data or missing values you're trying to predict, and so if you knew the whole population, you wouldn't need to do a regression—in fact, you'd be God.)

  The regression procedure basically says, "Here's the best-fit line *based on* the particular data you supplied, but you should check whether this line is truly meaningful and would apply to the broader population. So here's each coefficient, plus a *standard error* of that coefficient, plus a *t-statistic* for testing whether that coefficient is significantly different from zero, plus the associated *p-value*—now think about that!"

In Excel, a table with sample entries would look like this:

| | ***Coefficients*** | ***Standard Error*** | ***t-Stat*** | ***p-value*** | ***Lower 95%*** | ***Upper 95%*** |
|---|---|---|---|---|---|---|
| ***Intercept*** | 1.158 | 0.472 | 2.455 | 0.036 | 0.091 | 2.225 |
| **X *Variable*** | 0.964 | 0.080 | 12.088 | 7.24E-07 | 0.784 | 1.144 |

The columns tell you lots of juicy tidbits:

*Coefficients*: The best-fit line is $y = 0.964x + 1.158$. The second row contains the slope information, which is usually much more important than the intercept info.

*Standard error*: The slope 0.964 is just an estimate of the "true" slope of the best-fit line *for the whole population*, so the standard error tells you how good your estimate is. The smaller the standard error relative to the coefficient itself, the better the estimate. In this case, the standard error (0.080) is very small relative to the estimated slope (0.964), so the estimate is excellent.

t-*stat*: This tells you how many standard errors the coefficient is away from zero. The null hypothesis is *always* that each coefficient is zero. A high *t*-score is good (meaning that you'll be able to reject the null hypothesis and claim that the coefficient is significant). For instance, the *t*-stat for $\beta_0$ is 2.455, since 1.158 is 2.455 standard errors above zero.

p-*value*: This is the *p*-value associated with the *t*-statistic in the previous column, if you do a two-tailed test on each coefficient:

**$\beta_0$ intercept**

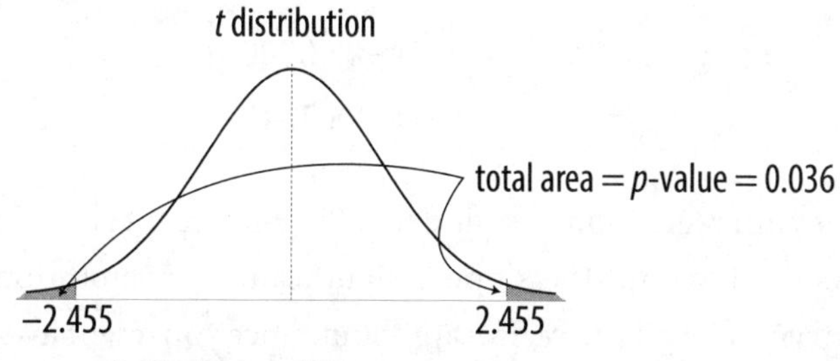


3.6% chance of being wrong
if you reject the null hypothesis for $\beta_0$

**$\beta_1$ slope**

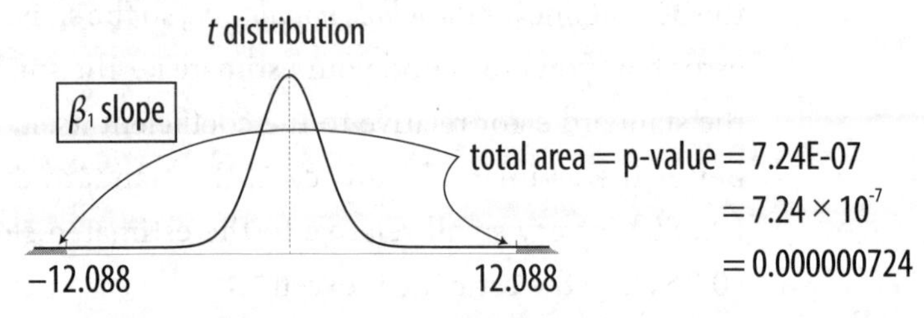


With these results, you can claim that the slope of the best-fit line is *not* zero with very high certainty (100% minus 0.0000724%, in fact!). You can claim that the intercept is not zero with reasonably high certainty too (100% minus 3.6%, or 96.4%).

Long story short—look for *tiny* p-*values,* such as 0.036 or 0.000000724. Those tell you that the coefficients you found are good.

Note: Output such as 7.24E−07 is very tiny. Convert to powers of 10 or decimals to see it clearly.

$$\begin{aligned} 7.24\mathrm{E}-07 &= 7.27\times 10^{-7} \\ &= 7.24\times 0.0000001 \\ &= 0.000000724 \end{aligned}$$

Sometimes in a regression, you find that the *y*-intercept isn't significant (*p*-value's too big) but the slope *is* significant. Pay attention to the slope—that's often all you really care about, since you might just want to know the *increase* in sales for every dollar of additional marketing spend.

## ESTIMATION VS. PREDICTION

So now that you have your beautiful line, you're all ready to use it.

$$y = 0.964x + 1.158$$

If $x = 20$, you predict...

$$y = 0.964(20) + 1.158$$
$$= 20.438$$

What you should recognize is that this specific ***prediction*** has two sources of error built into it:

1. Estimation error

   The coefficients of the line were only estimated from a sample, so the real best-fit line for the whole population might not be this exact line.

2. Additional random error

   Even around the theoretical best-fit line, there will be "noise," random deviation between the points and the line.

Here's another way to think about the issue.

Estimation: When $x = 20$, what do you expect the average value of $y$ to be across the whole population?

$$E(y \mid x = 20) = 0.964\,(20) + 1.158$$

Expected value of $y$, given $x = 20$, is 20.438.

Prediction: If you pluck *one* observation with $x = 20$ out of the whole population, what do you predict $y$ to be? You can't help it—you use the same equation:

Predicted $y = 0.964\,(20) + 1.158 = 20.438$.

The predicted single value of $y$ and the estimated mean value of $y$ (for a whole lot of observations) are the same number. But the error built into each is different.

Estimated mean value of $y = 20.438 \pm$ smaller error
(*only* the error that your line isn't quite right for the whole population)

Predicted specific value of $y = 20.438 \pm$ larger error
(estimation error plus random error, since the data doesn't all fall on even the best line)

## RESIDUALS

***Residual*** is another name for error or deviation—it's the vertical distance between some data point and the best-fit line.

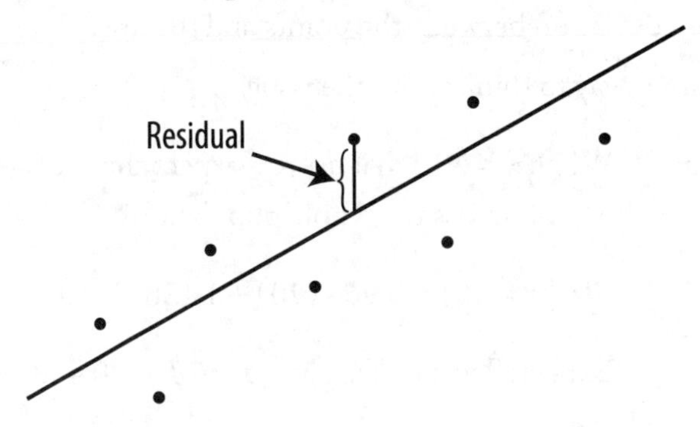


A residual can be positive or negative.

When you stress-test a regression to verify how robust it is, you look at the residuals. You want to make sure that they don't show a pattern:

- They should be independent of each other.
- They should be randomly scattered around zero, with more "small" residuals than "large" (ideally, they should form a normal distribution around zero).
- They shouldn't change in general size from one end of the plot to the other. For instance, the errors on the right side shouldn't be twice as big as those on the left side.

That last issue is more serious than it seems. If you run into it, you might have to rescale a variable to equalize errors. You'll learn a few tricks in class about how to rescale appropriately.

## MULTIPLE REGRESSION

Finally, you can have more than one predictor variable. For instance, you might try to predict sales of winter coats using both marketing spend and average temperature.

$$y = \beta_0 + \beta_1 x_1 + \beta_2 x_2$$

$x_1$ and $x_2$ are both predictor variables.
You still have just one $y$.

This is where art and science meet in real life—how many predictors to use, how they should be rescaled or combined. Remember a couple of key points about multiple regression, which is often the last topic in your first stats class (we say "first" because we know you're coming back for more):

1. Fewer Predictors = Usually Better

   You might think, "The more the merrier." In fact, the more *x*'s you throw in, the greater the chance that one of them gums up the whole works. Stick with fewer predictors, and your model will be cleaner and your life will be easier.

2. Independent Predictors = Always Better

   You don't want the predictors to be correlated. Here's the ideal scenario with two predictor variables:

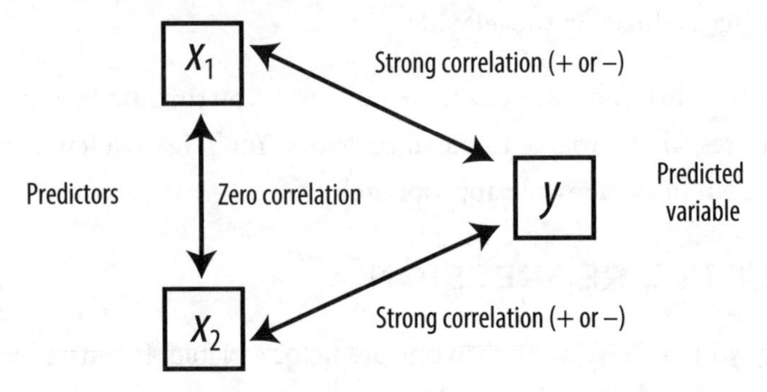


   If two predictor variables are strongly correlated, drop one of them or combine them sensibly somehow into a single index (add them, multiply them, or something). Then do your regression.

# Chapter 4: Accounting

## THE BIG PICTURE OF ACCOUNTING

Accounting is the standard way of keeping track of money in the business world. Accountants seem hard-wired to enjoy this stuff, even when they look haggard at tax time. So long as we get our refunds, we're happy to let them do the work.

Whether the subject is a regular business, a nonprofit, or even a single person (you), accounting answers these questions:

- Where is the money coming from and where is it going?
- How much do you own and how much do you owe? (Notice the difference a single letter makes between *own* and *owe*.)
- How are you doing financially *right now*?
- How is that financial picture changing over time?

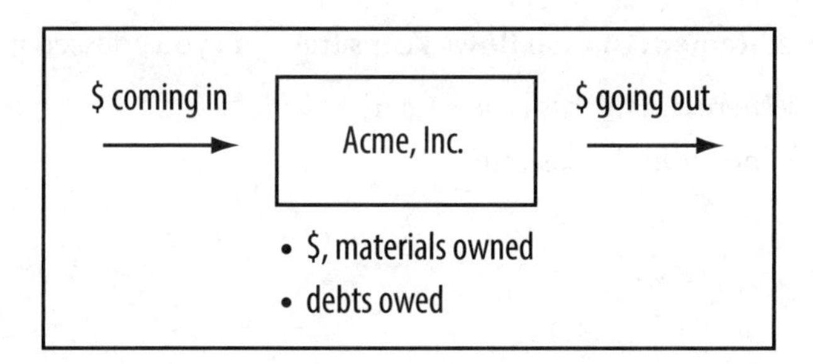

There are three types of accounting:

1. *Financial accounting* answers questions for outside investors and creditors. Standards are mandated by the government and are pretty complicated as a result.
2. *Managerial accounting* is for internal use. It builds off of financial accounting, adding customized metrics to help managers do their jobs well.
3. *Tax accounting* is for the IRS. Think April 15.

In b-school, you'll focus on types 1 and 2.

## THE BIG THREE FINANCIAL STATEMENTS

1. The **balance sheet** takes a *snapshot* on a particular date.
    - How much do I *own*?
    - How much do I *owe*?
    - How much is left *over*?
2. The **income statement** records *changes* to that picture over time.
    - How much did I *make*?
    - How much did I *spend*?
    - How much was left *over*?
    - What happened to what I *own* and *owe*?
3. The **statement of cash flows** keeps track of (you guessed it) cash.
    - Where did the *cash come from*?
    - Where did the *cash go*?

## BALANCE SHEET (WHICH HAS THE TERRIBLE ABBREVIATION B/S)

Let's start with the balance sheet, a very personal balance sheet: yours. Imagine that all you have in the world is $100 sitting in a checking account—and you borrowed $90 of that $100 from your Uncle Joey.

What you *own* = ***Assets*** (A) = $100 in the checking account

What you *owe* = ***Liabilities*** (L) = $90 loan from Uncle Joey (also known as debt)

What's *left over* = ***Shareholders' Equity*** (SE) = $10 that's yours (net worth)

What you *own* minus what you *owe* is what's left over:

$$A - L = SE$$

This fundamental equation drives the balance sheet.

Conceptually, this equation often gets rearranged a little. Think of *assets* as *resources*...

| **Assets** | |
|---|---|
| Checking Account | $100 |
| [There to help you operate] | |

and think of *liabilities* and *shareholders' equity* as *claims* on the assets.

| **Liabilities** | |
|---|---|
| Loan from Uncle Joey | $90 |
| [His claim on your assets] | |
| **Shareholders' Equity** | $10 |
| [Your residual claim on the assets, as you are the "shareholder" of you] | |

The claims ($90 and $10) are different. Uncle Joey has the *primary claim*. If your $100 goes down to $50 or up to $200, you still owe him $90. You, as the shareholder, have a *residual* or *secondary claim*. If your $100 goes down to $50, you're suddenly under water with $40 of "negative equity" (like all too many homeowners recently), but if the $100 goes to $200, your $10 of equity becomes $110. Shareholders take more risk than debtholders and get more potential upside as a result. That's the basic risk-reward tradeoff.

Since both liabilities and shareholders' equity are claims, go ahead and group them together on one side of the balance sheet.

$$A - L = SE \rightarrow A = L + SE$$

Now your balance sheet might look like this:

| Balance Sheet as of Today | | |
|---|---|---|
| **Assets** | Checking Account | $100 |
| | Total Assets | **$100** |
| **Liabilities** | Loan from Uncle Joey | $90 |
| | Total Liabilities | $90 |
| **Shareholders' Equity** | | $10 |
| | Total Liabilities & Share Equity | **$100** |

There are many different kinds of assets, liabilities, and shareholders' equity. You'll keep track of them in separate ***accounts*** with different names. Some of these accounts apply to a company more easily than to a person.

## ASSET ACCOUNTS

- *Cash*: Money in any kind of bank account.
- *Marketable securities*: Stocks and bonds you happen to have.
- *Accounts receivable*: Money that customers legally owe you.
- *Inventory*: Stuff you have to sell.
- *Advances to landlord*: If you prepaid your rent, then that's an asset from your point of view—you have squatter's rights for a while. That's worth something.
- *Other prepaid expenses*

All of the above are called ***current assets***—they're already equivalent to cash (the most liquid asset) or will be converted to cash within a year.

***Noncurrent assets*** turn into cash much more slowly:

- *Land*
- *Buildings* } Generally worth less now than when you bought
- *Equipment* } them. That decline in value is called depreciation.
- *Intangible* (but measurable) *assets*: Patents and brands are examples of intangible assets. To count patents as assets, you must have bought them from another company. Otherwise, they're not measurable (in dollars) from an accounting point of view. This is one of the limitations of accounting. A homegrown intangible asset that's never been bought or sold, such as the Coca-Cola brand, does not show up "on the books."

## LIABILITY ACCOUNTS

- *Accounts payable*: Money that you legally owe to someone.
- *Salaries payable*: Money that you legally owe to your employees.

Both of the above are examples of ***current liabilities***—debts you need to pay off within a year.

As you'd expect, there are also ***noncurrent liabilities***:

- *Long-term bonds*: Mortgages or other long-term debts.

## SHAREHOLDERS' EQUITY ACCOUNTS

These accounts apply better to a company than to a person.

- *Common stock*: The original *paid-in capital* by shareholders. In other words, this is the money the business got directly in exchange for shares of stock. It is *not* the current value of that stock in the market.
- *Retained earnings*: All the cumulative profits (or earnings) of the company that were left (or retained) in the company. The shareholders have rightful claim on the company's profits. They can either be paid those profits—as *dividends*—or they can leave them in the company as a kind of reinvestment. These reinvested profits are retained earnings. (Typically, management decides how much of the profits to pay out as dividends.)

Retained earnings is the account that balances the books—in other words, it makes $A = L + SE$ true. So compute everything else first. Once you know every other A, L, and SE account, you can solve for retained earnings.

## INCOME STATEMENT (I/S)

Unlike a balance sheet, which is a snapshot, the income statement is a movie. It records changes over time.

The movie you want to picture in your head is super-exciting: a pitcher of water filling up (good) or draining (bad). The following case is slightly simplified, but it captures the essence of the income statement.

1. Start of the movie: Jan 1.

The very first frame of the movie is a snapshot. This snapshot is the shareholders' equity on that day. Forget about whether this is $10 of assets, with no liabilities, or $100 of assets, with $90 owed to Uncle Joey.

2. Good changes through the year.

Over the year, you pour $50 into the pitcher. These $50 are called ***revenues***.

The water level rises.

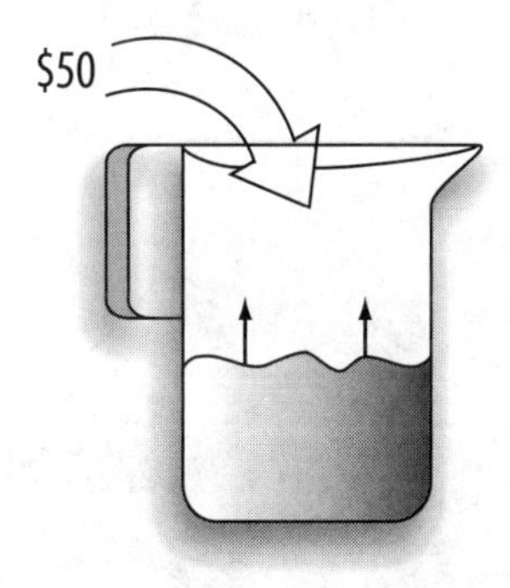

3. Bad changes through the year.

During the same period, you also spend $35 from the pitcher. These $35 are called ***expenses***.

The water level falls.

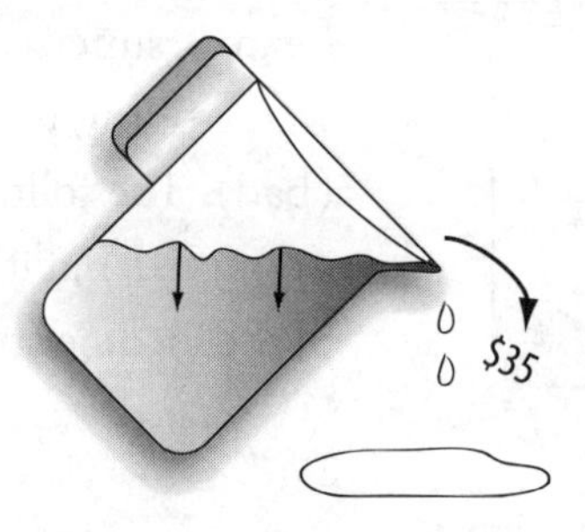


4. End of the movie: Dec 31.

How much is in the pitcher? Figure this out before you go on.

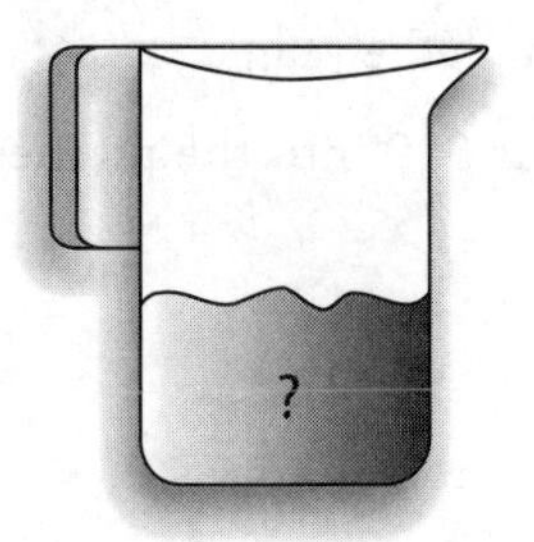


The math is straightforward:

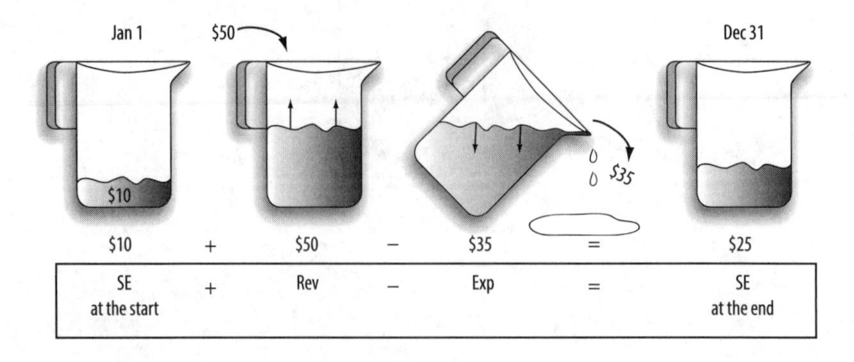

You often rearrange this to put revenues minus expenses on one side of the equation.

$$\$50 - \$35 = \$15$$

$$\text{Rev} - \text{Exp} = \text{Net Income (NI), or Earnings, or Profits}$$

That $15 is how much shareholders' equity (SE) changed (from $10 up to $25).

Now you have:

$$\text{Rev} - \text{Exp} = \text{NI} = \text{Change in SE}$$
$$\text{SE}_{\text{end}} - \text{SE}_{\text{start}}$$

$$\$50 - \$35 = \$15 = \$25 - \$10$$

This fundamental relationship drives the income statement. There are still a few questions to address.

*Question 1*: ***What time period is covered?*** (How long is the movie?)

*Answer 1*: ***A year or a quarter, usually.*** However, the year might not start on January 1. The *fiscal year* might start on July 1 or some other date.

*Question 2*: ***When do you "recognize" revenue?*** (When does a sale hit the income statement?)

*Answer 2-a*: ***When you get paid, fool.***

If you get paid $50 in cash today, then you "book" the revenue today. This method is called ***cash accounting***. This is how the IRS expects us to book our personal revenue (salary) when we pay our taxes for the year. What's the date on the check?

This method has attractive simplicity. However, it is also easy to manipulate, if you're a business.

*Answer 2-b*: ***When you finish most or all of the job and you get an asset*** (e.g., an IOU) ***in return.***

This method is called ***accrual accounting***. It means that you book the revenue when you do the work or deliver the goods—and the customer is obligated to pay you. It doesn't matter if that customer takes 30 days or 90 days to actually pay you. This is how all major public companies in the US track their revenue.

In the long run, cash accounting and accrual accounting are equivalent, since they only differ with respect to timing. Did GM book a particular sale in 1967 or 1968? It doesn't matter now. We're just glad they sold something.

*Question 3*: ***When do you recognize expenses?***

You have the same two possible answers:

*Answer 3-a, cash accounting*: ***When you pay the bill, fool.***

*Answer 3-b, accrual accounting*: ***When the related revenue is booked*** (matching principle) ***or when you get the benefits*** (period expense).

Say you're Walmart. If you sell a TV and book the sale now, you book the expense of purchasing that TV wholesale now (as "cost of goods sold" or COGS), even if you actually purchased that TV six months ago. That's called the matching principle. You "match" the sale to the expenditure. It works well when a sale and a related expense can be tied together.

Alternatively, some expenses (like rent) are not tied to particular revenues. In that case, you just spread the expense evenly over the period. If you prepay two years of rent for some reason, you book a chunk of that rent every quarter for the next two years.

By the way, we slightly simplified the income statement. We pretended that shareholders reinvested all the profits, but they might have gotten some dividends. If that's the case, then net income is not equal to the change in retained earnings exactly; you also have to add in the dividends to get net income.

| Net Income | = | Dividends | + | Change in Retained Earnings (in SE) |
|---|---|---|---|---|
| (All profits) | | (Profits distributed to shareholders) | | (Profits reinvested in the company) |

Equivalently, you could write that net income minus dividends equals the change in retained earnings.

## FROM REVENUES TO NET INCOME

An income statement from a real company tells a story. The exact details differ among industries and companies, but the overall gist is the same: you usually don't subtract off all the expenses in one fell swoop. Instead, you often subtract them off of revenues bit by bit, starting with the costs you can most directly attribute to the revenues, so that investors can understand the dynamics of the business better. Various intermediate steps in this "multiple-step income statement" are often computed on the road from revenues to net income. If nothing else, several big buckets of expenses are broken out on a single-step income statement so that investors can calculate those intermediate steps themselves and have some clue about how the business functions.

Here's a typical pathway, somewhat simplified:

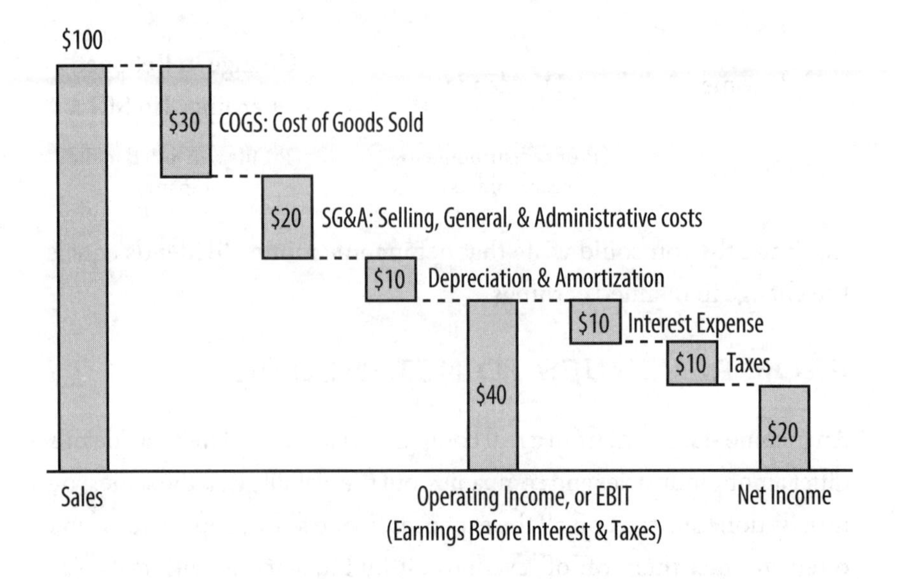


Revenues and expenses from ***going concerns*** and discontinued operations are generally separated from each other, as well as any unusual revenues or expenses—either one-time or from nonoperational activities, however those are defined. Often these unusual items are expressed as net *gains* or *losses* on the road to net income.

## THE BIG PICTURE IN PICTURES

Here's the balance sheet:

| | |
|---|---|
| Assets | $100 |
| Liabilities | $90 |
| + SE | $10 |

Top = Bottom

A = L + SE

Like a pitcher of water, the balance sheet changes as revenue pours in and expenses pour out.

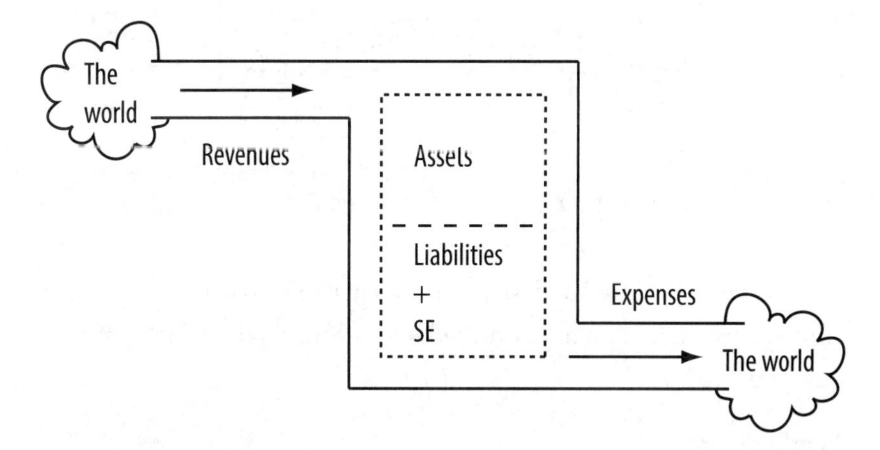


The income statement records those changes:

| Rev | – | Exp | = | NI | = | Div | + | Change in RE |
|---|---|---|---|---|---|---|---|---|
| | | | | Profit | | Dividends | | Retained Earnings ($RE_{end} - RE_{start}$) |
| \$50 | – | \$35 | = | \$15 | = | 0 | + | (\$25 – \$10) |

Be sure to grasp these relationships with simple numbers first. No matter how many zeros are tacked on, the concepts remain the same.

## TRANSACTIONS

A transaction in accounting world is just a flow of money from one "place" to another.

The places in question are accounts within any of the five big categories: Assets, Liabilities, Shareholders' Equity, Revenues, and Expenses. Just as before, an asset account tracks a specific kind of asset, a liability account tracks a specific kind of liability, and so on.

So a transaction is like a pipe that sucks some money out of one account and dumps it into another.

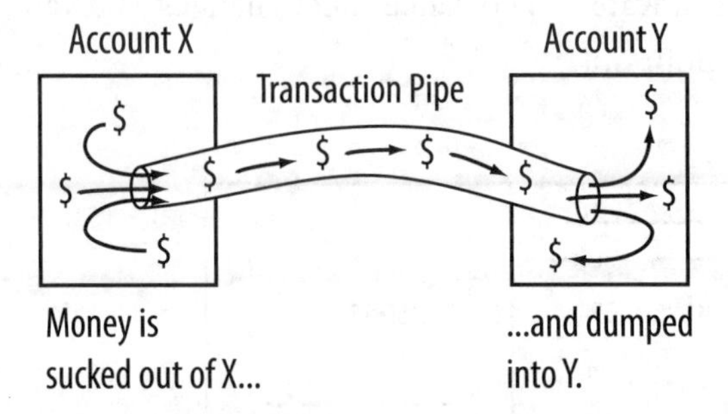


Money is never created or destroyed—it's just moved from one place to another. However much was sucked *out* of X has to equal the amount dumped *into* Y.

In other words:

$$\text{Outflow out of X} = \text{Inflow into Y}$$

In your personal life, you deal with this all the time. When you transfer $100 from checking to savings, you have an outflow of $100 from your checking account and an inflow of $100 into your savings account. Every pipe has two ends. When you withdraw $200 from the ATM, you have an outflow of $200 from some bank account and an inflow of $200 into your "cash in your pocket" account. Always look at outflows and inflows from the point of view of the accounts, not the pipe itself. $200 flowed *out* of the bank account and *into* your pocket.

Occasionally, one or both ends of the pipe splits, but the total outflows from one set of accounts still have to equal the inflows into the other set of accounts:

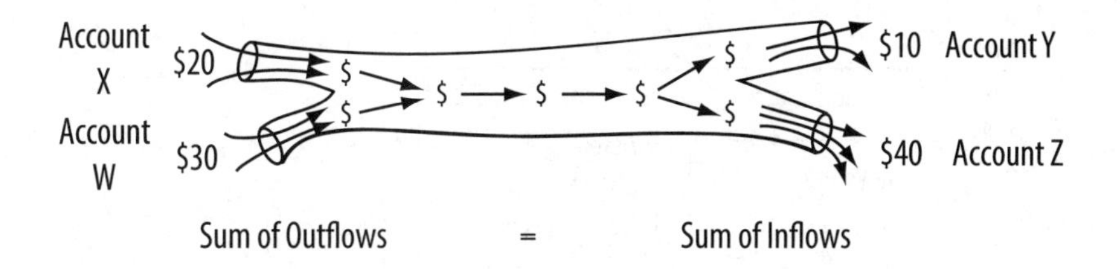


Take a look again at your original balance sheet. You have $100 in cash and you still owe Uncle Joey $90, so you have $10 in shareholders' equity. It's the start of the accounting period, so all revenue accounts (such as Salary) and expense accounts (such as Lunch) have been reset to $0. All that means is that so far this quarter or year, you haven't received any salary yet, and you haven't bought lunch yet.

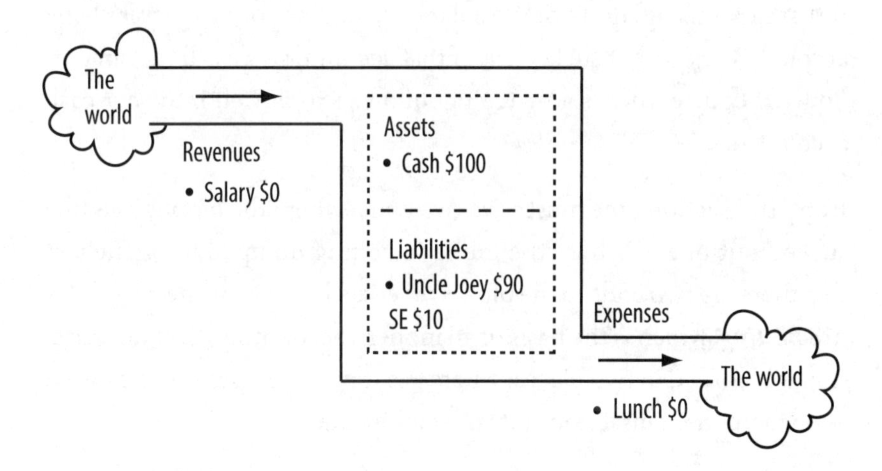


Now visualize some transactions.

1. You get paid $50 (cash) in salary.

The inflow is easy: $50 flowed *into* your cash account. Where did it come from? From a revenue account called Salary. So from the Salary account's point of view, that's an *outflow*.

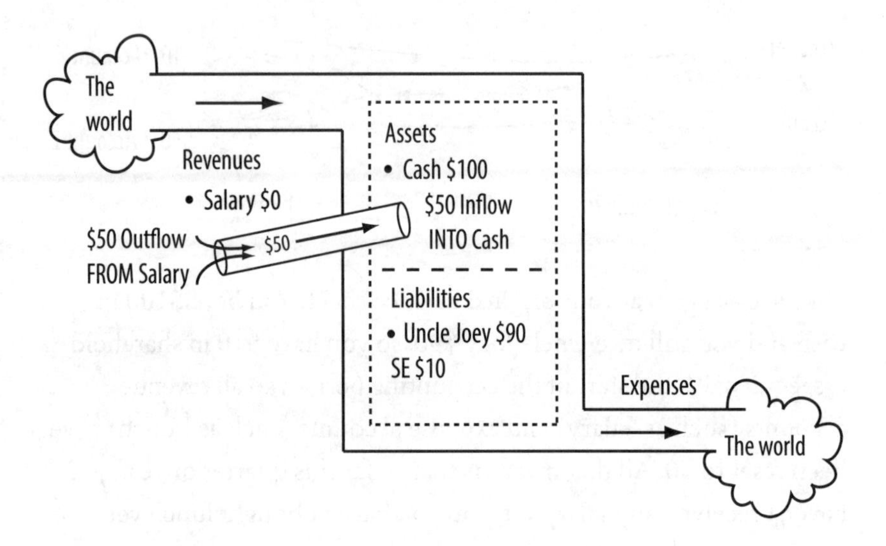


If that's confusing, think of the Salary account as your *employer's* bank account. The $50 got sucked *out* of that account, so we call that an outflow. Of course, there's a matching inflow: $50 flowed into *your* cash account.

Every transaction pipe must have two ends: where the money's getting sucked out of and where the money's getting dumped into. The fact that there are two ends to the pipe is reflected in the name ***double-entry accounting***, which is the basis of all modern accounting. Every transaction is recorded in two places: where the transaction pulls money from and where the transaction spits the money out.

Now, take some of your cash and do something with it. You're hungry but you're a little physically scared as well. So you'll do two more transactions to take care of your physical and psychological needs.

2. Pay $30 to your uncle. This calms you down.

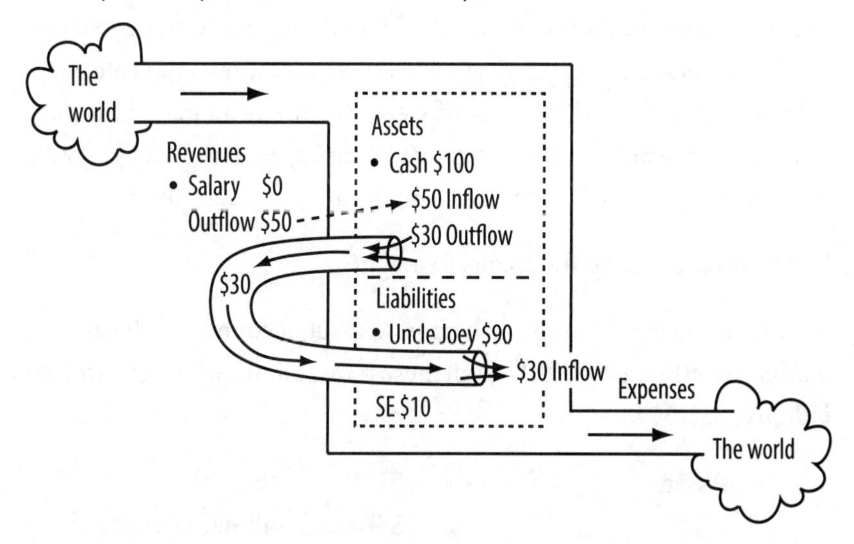


3. Spend $35 on lunch to assuage your hunger. Given your uncle's propensity toward violence, this may not be the wisest use of funds, but you must record the transaction properly nonetheless:

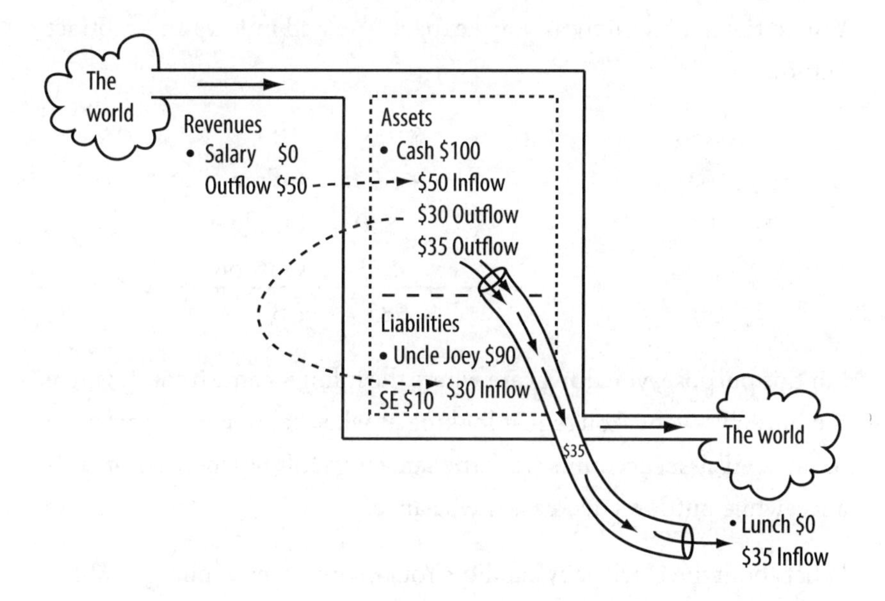

Notice that *revenue* accounts typically have *outflows* (from *their* point of view). These outflows are matched by *inflows* into other accounts, usually assets such as cash. Meanwhile, *expense* accounts typically have *inflows* (again, from *their* point of view). To the restaurant, your $35 went *into* some account. That inflow into an expense is matched by an outflow from some other account, usually an asset or a liability.

It's important to keep the changes straight.

Start with the cash account, which is an asset account. The ***beginning balance***, or BB, was $100. After all these transactions, what's the ***ending balance***, or EB?

| Assets | • Cash | $100 | BB |
|---|---|---|---|
| | | $50 | Inflow |
| | | $30 | Outflow |
| | | $35 | Outflow |
| | | $? | EB? |

You do the math exactly as you'd expect. You add inflows and subtract outflows.

| Assets | • Cash | $100 | BB |
|---|---|---|---|
| | | + $50 | Inflow |
| | | – $30 | Outflow |
| | | – $35 | Outflow |
| | | $85 | EB |

For this purpose, you don't care where the inflows came from or where the outflows went. You're just looking at what happened to *your* cash balance. All asset accounts work the same way. Inflows increase the balance, while outflows decrease the balance.

What about the Uncle Joey liability? You originally owed him $90. What do you owe him now?

| | | | |
|---|---|---|---|
| Liability | • Uncle Joey | $90 | BB |
| | | $30 | Inflow |
| | | $? | EB? |

Think of any liability as a credit card. What happens to your balance when you pay some money to your issuer? Your balance goes *down*.

| | | | |
|---|---|---|---|
| Liability | • Uncle Joey | $90 | BB |
| | | – $30 | Inflow |
| | | $60 | EB |

If you like, you can think of liabilities as "negative money." In fact, accounting predates the theory of negative numbers, and some of the first mathematical explanations of negative numbers referred to positive numbers as fortunes (assets) and to negative numbers as debts.

When you put $30 of normal, positive money against negative $90, you get negative $60. You're bringing the debt closer to zero:

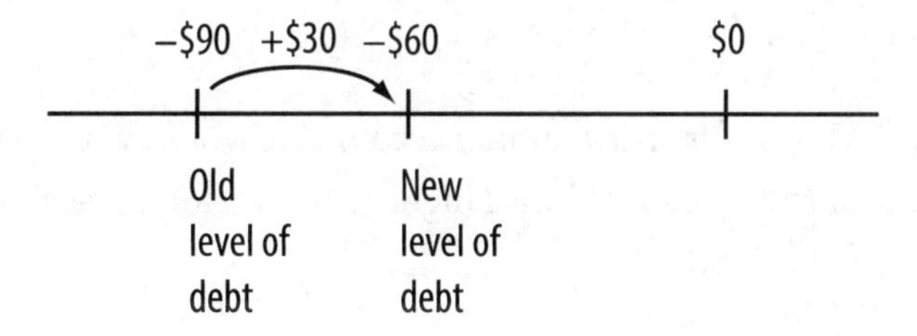


It's probably easiest, though, just to think of what happens to your credit card bill. Inflows of money *decrease* the balance of any liability. Outflows of money (using your credit card) *increase* the balance. As you'd expect, this is the exact opposite of what happens with assets.

How about revenues and expenses?

Since revenue accounts usually just have outflows, that's what we total up over any period of time. By the way, an inflow back into a revenue account would not be impossible, but it would indicate something weird:

your boss overpaid you and you have to return some of the money. In a business context, a refund to a customer would fall into this category.

Conversely, expense accounts usually just have inflows into them. An outflow back upstream from an expense account indicates something weird: a refund paid back to you from a vendor, for instance.

## THE EVIL LANGUAGE OF ACCOUNTING

Unfortunately for all of us, the accounting world refers to outflows and inflows by crazy, counterintuitive terms. In fact, these terms mean *precisely the opposite* in accounting from what they mean in your everyday life.

Let's say you have an inflow of $50 into your checking account. What would you call that, a credit or a debit?

You'd *think* it was a credit, right? But no, in accounting land, that's called a ***debit.*** If you *debit* a cash account $50, you are putting $50 *into* it!

It's insane, but true.

Likewise, if you ***credit*** an account in accounting world, you are sucking money *out* of that account! This language is a historical accident that we all have to live with.

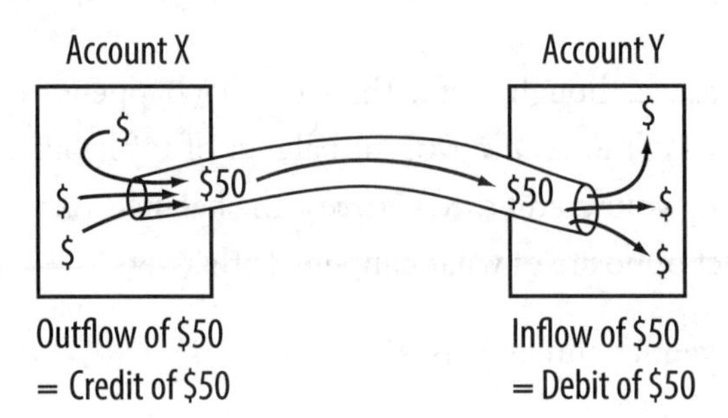


Amazingly, accounting textbooks and professors seem to ignore or gloss over the fact that this terminology is completely bass-ackwards. What some of them do, instead, is tell you to forget what these terms *mean,*

and just memorize a bunch of rules about what credits and debits *do to* different types of accounts.

That's a bad strategy.

A much better strategy is to think in terms of actual movements of money, as you've been doing, and just accept the fact that the ends of the pipe have been misnamed in "accounting Latin."

An *outflow* of money out of an account is always a *credit.*

Outflow = Credit

An *inflow* of money into an account is always a *debit.*

Inflow = Debit

If you just memorize these reversed terms, then you can actually use your common sense to figure out what's happening in the language of credits and debits. Just remember: "*credit out, debit in.*"

Look again at the three transactions in the big picture, replacing the language.

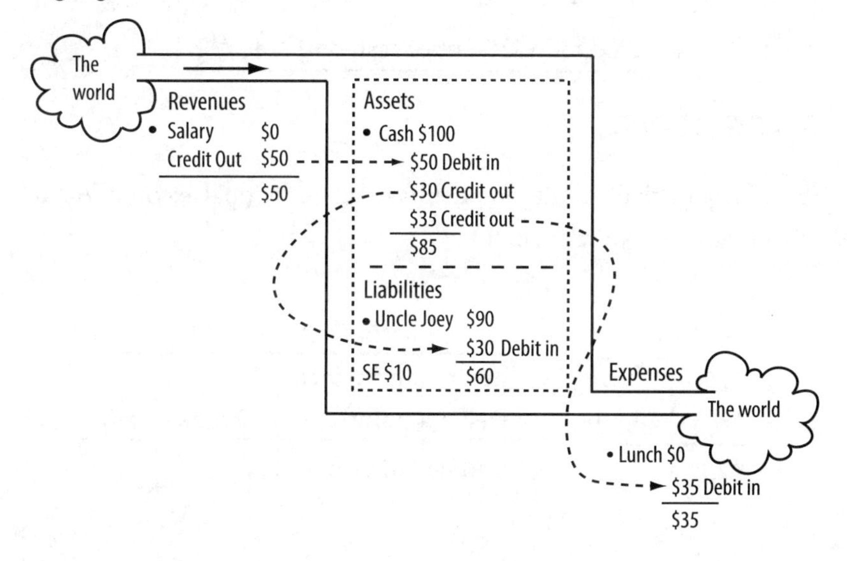

If you just consider the $30 payment to Joey, you can see a couple of key accounting tools in action.

## JOURNAL ENTRY

Think of the journal as a diary: an entry records a transaction in a very standardized way. The pattern is always as follows:

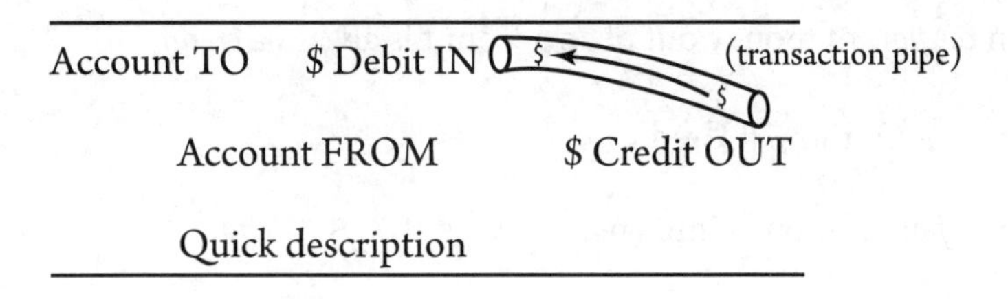


Why they always put the destination before the starting point of the money is another mystery. Plus, you have to imagine the money flowing left. But you can deal. Here's the Joey payment as a journal entry:

| | | |
|---|---|---|
| Joey's loan | $30 | |
| | Cash | $30 |

Paid Joey $30 out of cash on 12/15/11.

## T-ACCOUNTS

These keep track of credits and debits for an individual account. Again, consider the $30 Joey payment.

| Cash | |
|---|---|
| $100 | |
| | $30 |
| $70 | |

| Any Asset Account | |
|---|---|
| Beginning Balance<br>Debits (if any) | Credits (if any) |
| Ending Balance | |

You add debits and subtract credits.

It would be easier if we just wrote:

| Asset | |
|---|---|
| Beginning Balance<br>Inflows | Outflows |
| Ending Balance | |

You add inflows and subtract outflows. This is what you're actually doing with debits and credits—don't lose sight of that.

Balances for liabilities go on the right:

| Uncle Joey | |
|---|---|
| | $90 |
| $30 | |
| | $60 |

| Any Liability Account | |
|---|---|
| | Beginning Balance |
| Debits (if any) | Credits (if any) |
| | Ending Balance |

For liabilities, you add credits (outflows) and subtract debits (inflows). After all, you want your liabilities to go down.

In all cases, credits go on the right and debits go on the left. This is frequently cited as if it were something truly deep and explanatory. Yes, it's good to know, but it's no substitute for knowing what credits and debits actually are (outflows and inflows). Of course, sometimes you'll solve a simple problem mechanically. However, unless you understand that a credit is an outflow of money from an account, while a debit is an inflow of money to an account, you'll never be able to reason through a complicated scenario correctly.

We harp on this point because this is where your accounting professor will typically fail to grasp the difficulty. Again, accountants are a special breed, whom we must love just as they are.

## TYING REVENUES AND EXPENSES INTO THE BALANCE SHEET

Consider the $50 salary payment for a minute.

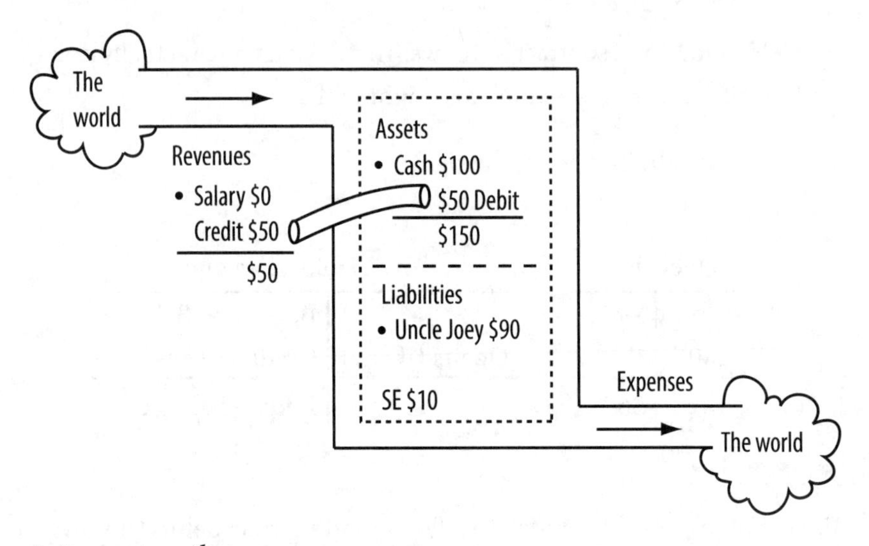


Is it now true that

| Assets | = | Liabilities | + | SE? |
|---|---|---|---|---|
| $150 | = | $90 | + | $10? |

Nope! ~~$150 = $100~~

Something's definitely wrong with this picture. Here's what's going on: you now have $150 in cash. And you still only owe $90 to Uncle Joey. So you must have $150 – $90 = $60 in shareholders' equity, not $10.

This means that this revenue account, which now shows $50, must actually be a shareholders' equity account for A = L + SE to still work.

| | | |
|---|---|---|
| Assets | $100 | After the transaction, |
| • Cash | $50 | |
| | $150 | A = $150 |
| Liabilities | | |
| • Uncle Joey | $90 | L = $90 |
| SE | | |
| • Retained Earnings | $10 | |
| • Salary | $50 | SE = $60 |
| Total SE = | $60 | L + SE = $150 |

The same thing is true of expense accounts. Technically, both revenues and expenses are "temporary" shareholders' equity accounts.

At the end of any accounting period, you have to zero out every revenue or expense account's balance by moving money to or from that catch-all SE account, retained earnings—the one that captures all reinvested profits of the company since it was founded.

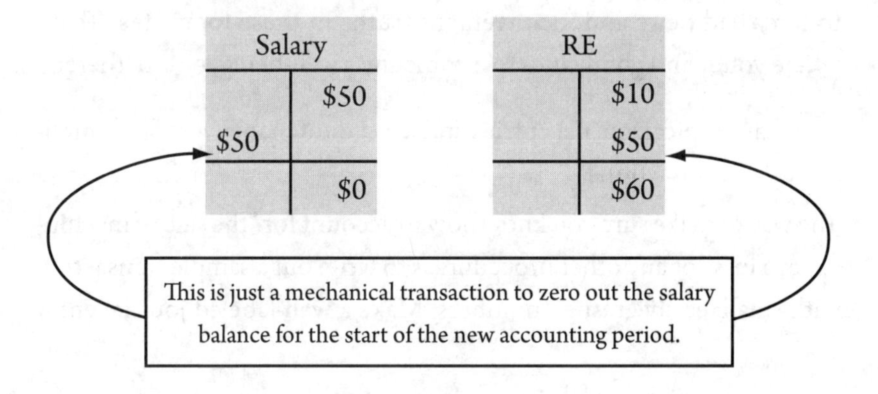

## FINAL THOUGHTS ON FINANCIAL ACCOUNTING

1. The Statement of Cash Flows is a beautiful thing. Really? Yes.

You'll learn how to build and interpret the SoCF in your accounting class. If you've understood the previous material, you'll be fine. Just realize that the SoCF sheds wonderful light on the movement of cash. In fact, the content of the SoCF overlaps significantly with that of the B/S and the I/S. In theory, with sufficiently detailed B/S's and I/S's, the SoCF is redundant. However, that's not the point. The purpose of the SoCF is to reveal how cash is flowing to and from three principal sorts of activities:

1. Operations (cash from customers, less cash paid to suppliers, employees, etc.)
2. Investing (purchase and sale of long-term assets)
3. Financing (cash to and from investors, either shareholders or bondholders)

Without a careful handle on these cashflows, a firm might look good on its balance sheet and income statement but run out of dough midyear.

When you're doing an accounting case, you're often doing *forensics*: what the heck happened to this company over the last six months? Word to the wise—look at the SoCF. The other good place to look for buried treasure is in the "Notes to the Financial Statements." Companies love to bury bad news and inconvenient truths in these footnotes. That's where you'll find good clues to a company's well-being, or lack thereof.

2. Learn typical journal entries and T-account maneuvers for complicated transactions.

The way to make sure you know how to account for "the sale of machinery at a loss" or any other procedure is to write out a sample transaction with small, easily grasped numbers. Make a well-labeled journal entry on your review sheet, whether you can take such a sheet into your exam or not. Likewise, create sample T-accounts for funky accounts.

For example, Allowance for Uncollectibles is a "contra-asset" account, a kind of liability account attached directly to the Accounts Receivable asset account to track money you won't ever collect from your deadbeat customers. Here is the notation on the review sheet one of us was able to create and use on the final:

| Allow for Unc (XA) | |
|---|---|
| | BB |
| write off | bad debt expense |
| | EB |

The important thing is not that you know what these terms mean right now. Rather, just remember to come up with meaningful notations for typical credits and debits in tricky accounts. This way, you'll know how to use those accounts.

## A FEW POINTS ON MANAGERIAL ACCOUNTING

Managerial accounting (also known as cost accounting) is a separate subject from financial accounting, which is what has been covered so far. It is not as quantitatively or conceptually hard, thank goodness, so we'll say much less about it here.

A couple of the key questions in this field:

1. How do you properly allocate costs to different product lines, so you can compare profitability?
2. How do you track performance against a budget?

To address the first question, you first have to distinguish ***variable costs*** from ***fixed costs***. Variable costs change with the level of production. For instance, the ***direct cost*** of the materials out of which a product is made is a variable cost. Some ***overhead costs*** are also variable. In contrast, a *fixed cost* doesn't vary over wide ranges of production. For example, the CEO's salary is a fixed cost, at least in the short run.

The interesting issue is how to allocate these fixed costs and other costs that aren't natually attributed to different product lines. In theory, you go to town with ***activity-based costing*** and track everyone's activities as they relate to your various products. There will always be some residual fixed costs, though. The typical approach is to compute a ***contribution margin*** for each product.

| | |
|---|---|
| Revenues from Sales | $100 M |
| Variable Cost of Goods Sold | – $40 M |
| Contribution Margin | $60 M |

The contribution margin then "contributes" to covering the fixed costs. Contribution margins are very proprietary pieces of information because, together with fixed costs, you can compute the ***breakeven point***, the number of units sold below which the firm is losing money and above which it's making money. At this point, the firm "breaks even" with zero profit.

Say that a particular product sells for $200. The variable COGS is $80, so the contribution margin per unit is $120. This means that every unit sold contributes $120 toward covering whatever fixed costs this product is responsible for. Say those fixed costs are $12 million. Then the company must sell $x$ units to break even, where $x$ satisfies:

$$\underbrace{(120)(x \text{ units})}_{\text{total contribution margin}} = \underbrace{\$12 \text{ million}}_{\text{fixed costs}}$$

$$x = \frac{12 \text{ million}}{\$120} = 100{,}000 \text{ units}$$

More generally, the breakeven point (BE) is given by:

$$BE = \frac{\text{Fixed Costs}}{\text{Contribution Margin per Unit}}$$

Graphically, the relationship between profit and units sold can be seen this way:

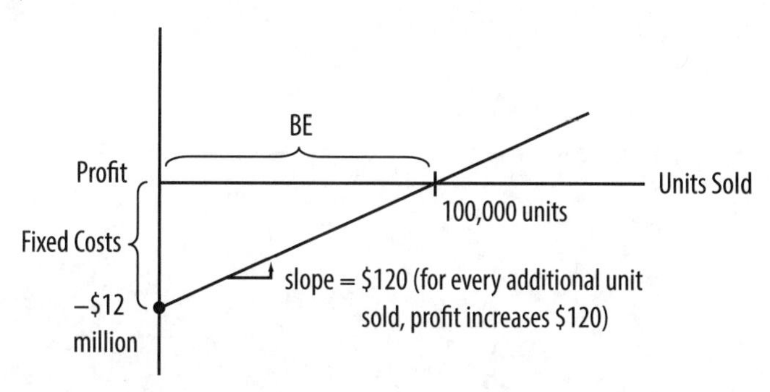


As for managing performance against a budget, the game is all about computing ***variances***—not in the statistical sense, but in the more general sense of "differences." The idea is this: if you sell $150,000 more than you predicted, how much of that difference was because you sold more or fewer *units* and how much was because you sold at a different *price* than you predicted? Likewise, if your costs are not what you budgeted, which department should be blamed—if at all?

The computations are not difficult; it's just a matter of learning the terms and keeping good track.

# Chapter 5: Finance

Whether it's "FIGH-nance" or "fi-NANCE," this stuff is at the heart of business. Learn it well (or at least well enough to get through exams).

## BIRD IN THE HAND

The most fundamental finance concept is the ***time value of money***. Here it is, in a nutshell:

| $100 Today | = | More than $100 Tomorrow |
|---|---|---|

Why is this true? Two main reasons:

1. ***Delayed gratification.*** Say you owe Jill $100. If you pay her right now, she can spend that $100 right away, so if you're going to make her wait till tomorrow, you need to give her a little more than $100.
2. ***Risk.*** What if between today and tomorrow, you take off for Austria, leaving Jill high and dry? She trusts you, but she still needs a little extra incentive to wait and take the risk you might leave town. Plus, because of inflation, a dollar might not buy as much tomorrow as it does today.

Together, delayed gratification and risk mean that you have a choice: pay Jill $100 now or pay her more than $100 later.

A way to visualize this equivalence is with column charts:

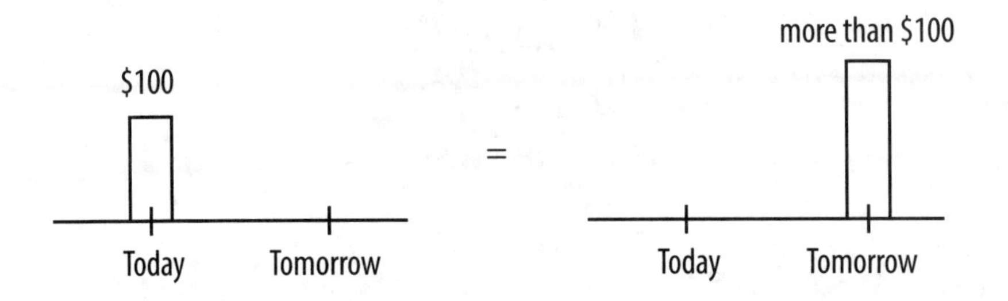


As you move a column of money (called a ***cash flow***) to the right and thus into the future, it has to grow to stay equivalent. That growth is the ***interest*** you have to pay Jill to accept the delay and to take the risk that you might not pay her.

Now try slightly larger specific numbers and change the time frame to a year:

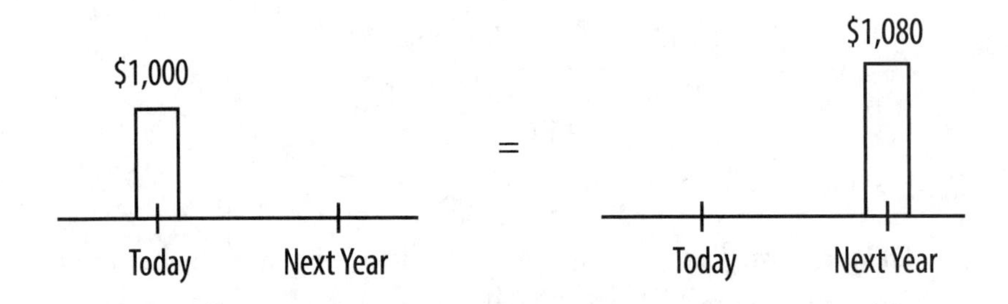


In words, here's the situation: you owe Jill $1,000 today. She tells you that you can either pay her the grand right now, or you can pay her $1,080 in exactly one year.

You took either the GMAT or the GRE, so you know something about the math here. By asking for $1,080 next year, Jill is charging you an 8% annual ***interest rate***.

$$\$1,080 = \underbrace{\$1,000}_{\text{(the principal)}} + \underbrace{8\% \text{ of } \$1,000}_{\text{(the interest)}}$$

$$= \$1{,}000 + (0.08)(1{,}000)$$

$$(\text{since } 8\% = \frac{8}{100} = 0.08)$$

To go even further, you could pull out the $1,000 "common factor" on the right, which leaves behind a 1 and a 0.08.

$$\$1{,}080 - \$1{,}000 + (0.08)(\$1{,}000)$$

$$= \$1{,}000(1 + 0.08) = \$1{,}000(1.08)$$

Here's a key bit of math to get comfortable with:

| Adding 8% to some amount of money | = | Multiplying that amount of money by 1.08 |
|---|---|---|

Likewise, adding 13% to some amount is the same as multiplying that amount by 1.13. You can generalize if you've expressed the interest rate as a decimal $r$ (like 0.08 or 0.13) and you have $1,000 today:

| Next year's equivalent amount of cash | = | $\$1{,}000\ (1 + r)$ |
|---|---|---|

If you like percents better, then there's just a slight difference in the formula. If you write the interest rate as $x$ *percent* ($x$%), then just write in $\frac{x}{100}$:

| Next year's equivalent amount of cash | = | $\$1{,}000\ (1 + \frac{x}{100})$ |
|---|---|---|

After all, 1.08 can be written as $1 + 8/100$. Just keep track of whether your variable is already a decimal (like $r$ above). If the variable isn't a decimal already, you need to divide it by 100 (like $x$ above).

If $P$ is today's principal and $r$ is the annual interest rate (as a decimal), you can see the column equivalence this way:

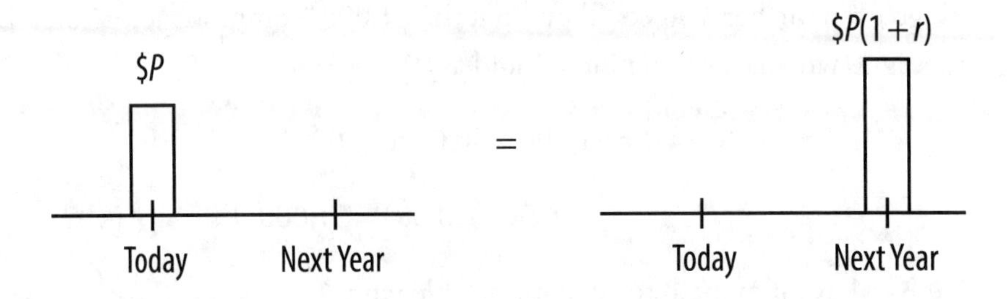


As a shorthand, put these pictures together:

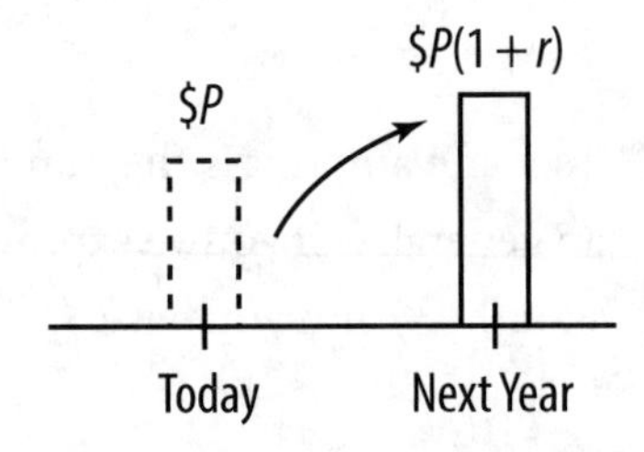


The way this picture is drawn indicates the way we've been thinking: today's money earns interest and becomes a bigger amount of money next year.

## THE POWER OF COMPOUNDING

What if you let the money sit for *two* years? Then you have something called ***compounding***. You earn "interest on the interest," so in the second year, you earn a little more interest than in the first year.

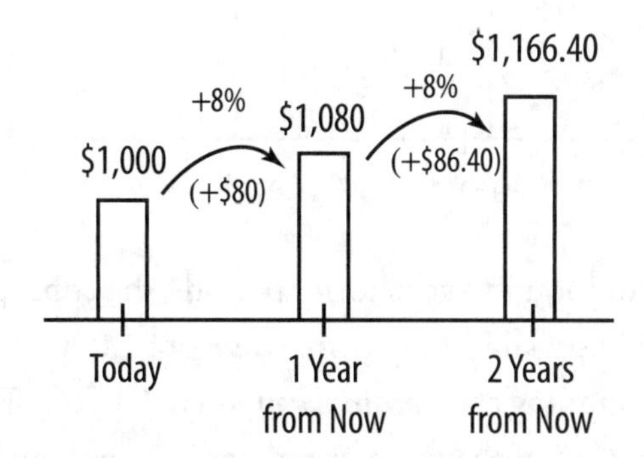


Notice that you're not *adding* the same amount of money each year. However, you are doing the same multiplication each year: both times, you're multiplying by 1.08.

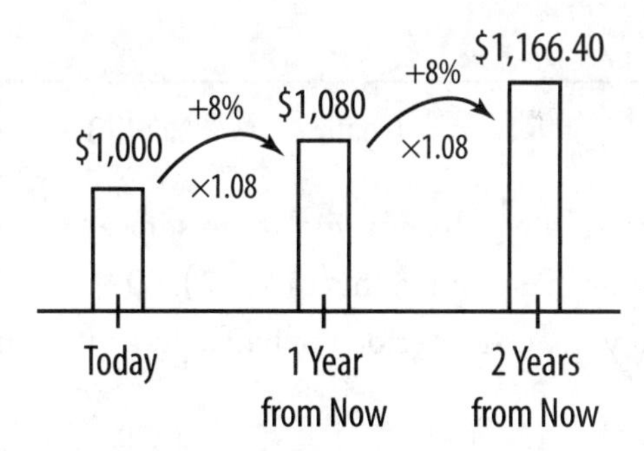


This means that you can write the final amount this way:

$$\begin{aligned}\$1{,}166.40 &= \$1{,}080(1.08)\\ &= \$1{,}000(1.08)(1.08)\\ &= \$1{,}000(1.08)^2\end{aligned}$$

Compounding over several years is the same as multiplying by a number bigger than 1 several times:

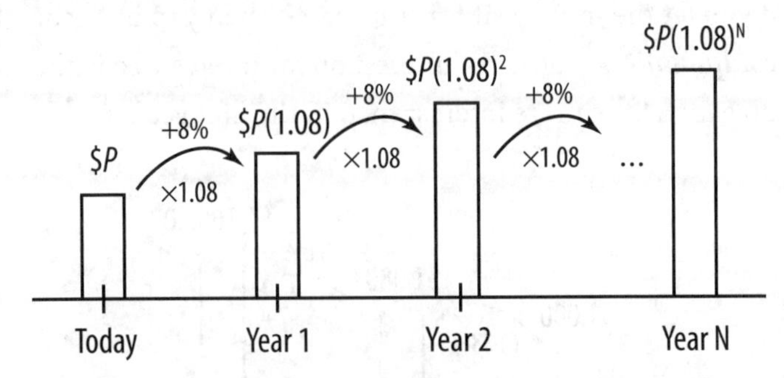


If you have to compound over shorter periods, the concept works similarly. Say you'd like to find the ***future value*** of $1,000 after one year of monthly compounding at a ***nominal annual rate*** of 6%. Then you don't have just one period of a year—your period is one month, so you have 12 periods in that year.

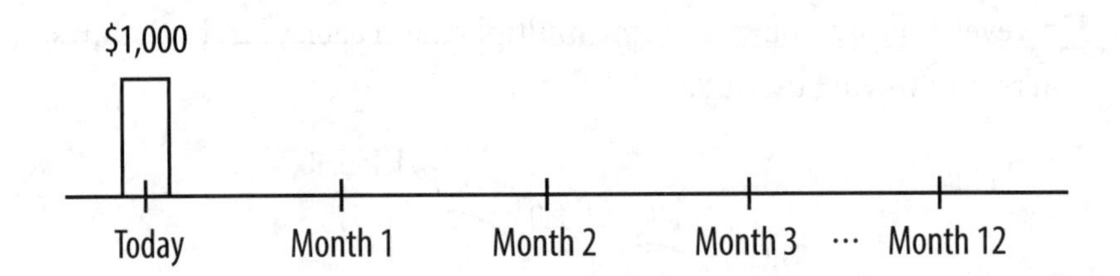


The period interest rate is the nominal interest rate (6%) divided by the number of periods in a year (in this case, 12). 60 ÷ 12 = 0.5%, or 0.005 as a decimal. So you get 12 periods of half a percent interest:

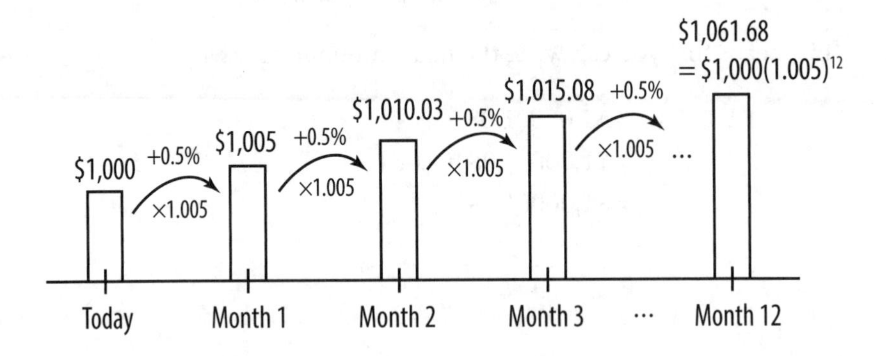

This is slightly better than waiting till the end of the year and getting 6% all at once, since that non-compounded 6% is only calculated on the $1,000 principal. In contrast, when you're compounding along the way, you're getting little bits of interest on the interest, yielding you an extra $1.68 in this case. This is not a lot of money, but imagine a principal of $1 million. Or $1 billion. Then compound interest becomes real money you care about.

A nominal annual interest rate of 6%, compounded monthly, is equivalent to an ***effective annual interest rate*** of 6.168%. Your credit card company knows all about this, of course.

## BRING IT BACK

So far, we've been thinking about the growth of money *into the future*. For instance, $1,000 today turns into $1,080 next year, at an annual interest rate of 8% (a ***simple*** rate, not compounded).

Now let's flip the script.

If $1,000 today is equivalent to $1,080 next year, doesn't that also mean that $1,080 next year is equivalent to $1,000 today?

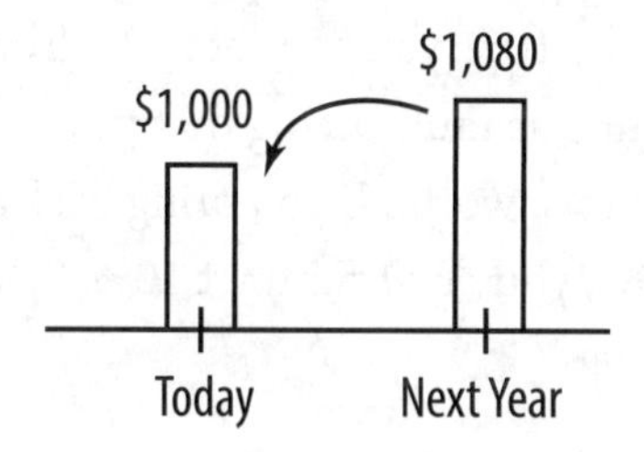


The rule is simple. To go forward in time, you multiply by 1.08. So to go back in time, you divide by 1.08.

| Today | | Next Year |
|---|---|---|
| $1,000 | $\xrightarrow{\times 1.08}$ | $1,080 = $1,000(1.08) |
| $1,000 | $\xleftarrow{\div 1.08}$ | $1,080 |

$$\$1{,}000 = \frac{\$1{,}080}{1.08}$$

In finance, you are often pulling a future cash flow back in time to the present, in order to figure out what it's worth today. To do this pull-back, you *divide* the cash flow by $1 + r$, where $\boldsymbol{r}$ is called the ***discount rate***.

For instance, a cash flow of $500 one year from now, discounted at 10%, is only worth $454.54 today:

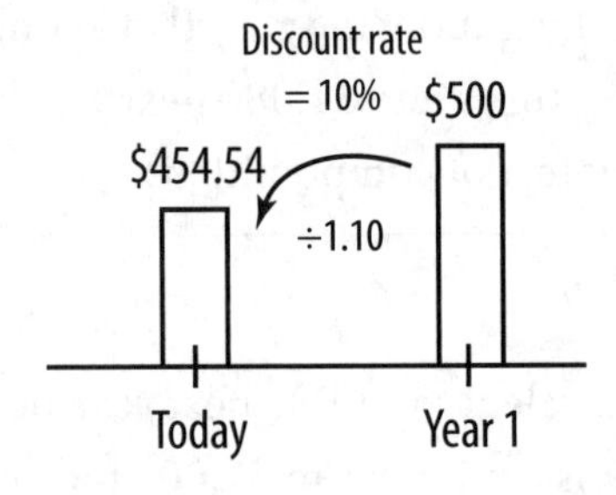


The word *discount* can be misleading. In the retail world, if you discount the price of a sweater by 10%, you subtract 10% from the original price, which is the same thing as multiplying that price by 90% or 0.9. Contrast that with the finance world. If you bring back a future cash flow at a discount rate of 10%, you *don't* subtract 10%. Instead, you divide the cash flow by 1.10.

Your aim is to solve this puzzle—what is the *current* dollar figure that, if you added 10% to *it*, would give you $500 (or whatever the future cash flow is)? The answer is $\frac{\$500}{1.10} = \$454.54$.

If you pull a future cash flow back in time across multiple periods, you do the reverse of compounding. Rather than multiply repeatedly, you divide repeatedly.

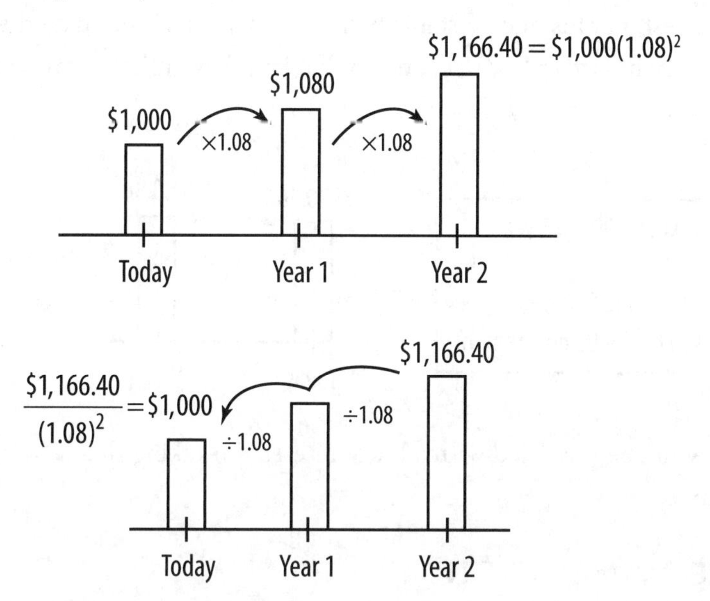


## RISK AND REWARD

The *higher* the discount rate, the *less* some promised future cash flow is worth today.

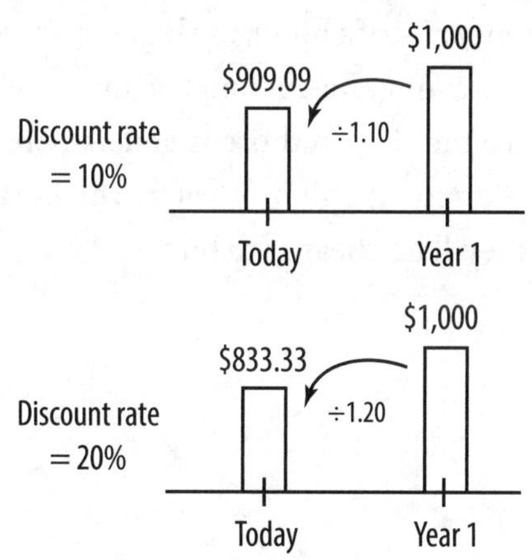

These different ***present values*** reflect a difference in perceived risk. The higher the discount rate, the riskier the investment. A low-risk investment has a low discount rate—in other words, it pays a low effective rate of interest. For instance, a stable bank with deposits insured by the US government is very low risk, and so that bank's promise to pay $1,000 in a year is worth very close to $1,000 now.

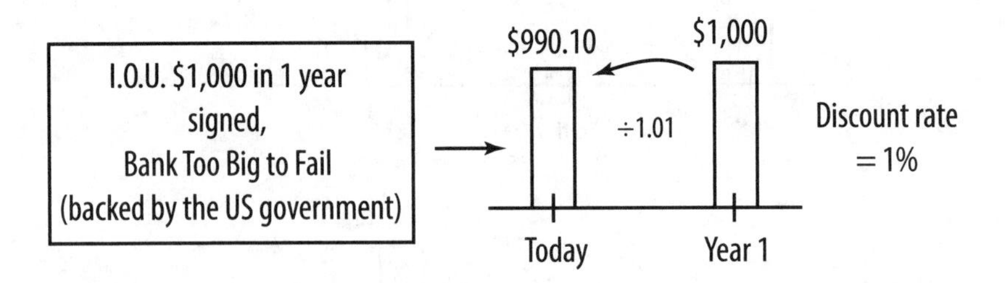


You won't be rewarded with a lot of interest for taking such a low-risk gamble.

## THE NAME'S BOND...

If the I.O.U. is traded publicly on Wall Street, you can "impute" or figure out the effective discount rate from the marketplace. This kind of I.O.U. is often called a ***bond***.

The lower the current ***price*** of a bond, the higher its ***yield***, a.k.a. effective interest rate, a.k.a. discount rate. A company might sell some I.O.U.'s to the public, but since the risk on those is generally higher than that of the bank bond, the discount will be greater. The I.O.U.'s will be worth less right now, so they'll be cheaper to buy.

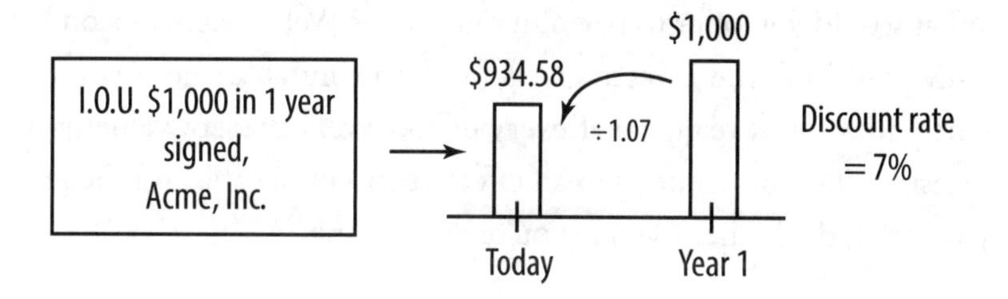


| Borrower/Issuer | Current Value of 1-Year $1,000 Bond | Discount Rate, or Yield |
|---|---|---|
| Bank Too Big to Fail | $990.10 | 1% |
| Acme, Inc. | $934.58 | 7% |
| Uncle Joey | $833.33 | 20% |

The riskier the borrower, the greater return you'll demand on your money, so the yield goes up. You hand over less right now to that risky borrower in return for the promise of $1,000 in a year.

## MULTIPLE CASH FLOWS

If an investment promises you more than one cash flow at different points in time, then you bring each one back to today by discounting, then add up all the results. Say that Uncle Joey has an "investment vehicle" to tell you about:

What would you offer to pay him right now? Well, it depends on how risky you think the investment is (and how much or little you trust Uncle Joey). Just realize that every offer—every present value of that investment—corresponds to an effective discount rate, and the lower your offer, the higher the rate you're demanding.

Let's say that you think that this investment ought to be paying you 15% all the way through. Here's how to figure out your offer:

1. Draw a picture of the proposed future cash flows.

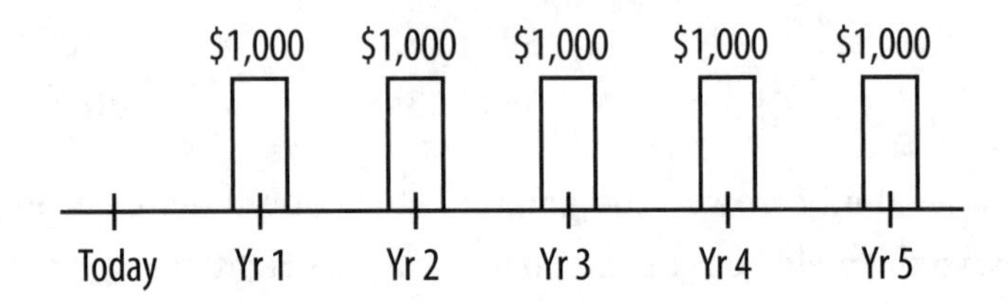


2. Bring each cash flow back to today. Since the discount rate is 15%, we need to divide by 1.15 (= 1 + $r$, where $r = 15\% = \frac{15}{100} = 0.15$).

   We have to divide by 1.15 for every year between now and the future cash flow. That's the same as squaring, cubing, etc. 1.15 and dividing by the result.

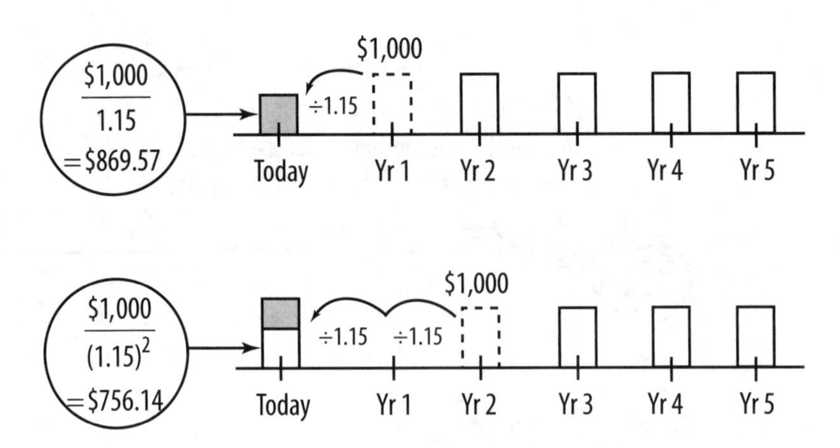

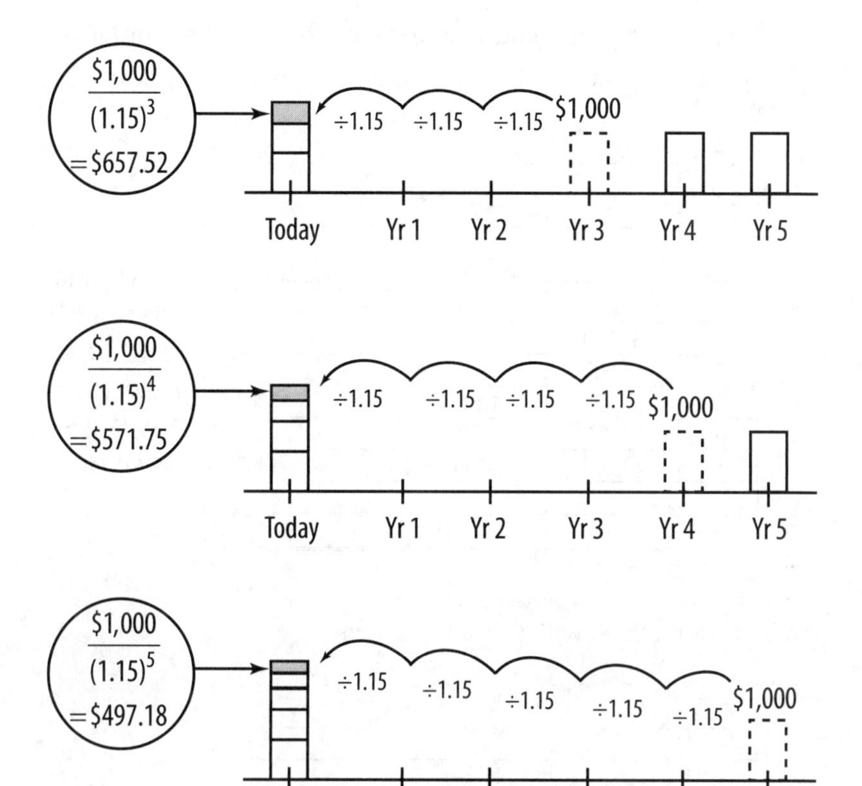


3. Now add up all those present values to get the ***net present value***, or ***NPV***, of the investment.

$$\text{NPV} = \frac{\$1{,}000}{1.15} = \frac{\$1{,}000}{(1.15)^2} + \frac{\$1{,}000}{(1.15)^3} + \frac{\$1{,}000}{(1.15)^4} + \frac{\$1{,}000}{(1.15)^5}$$

$$= \$869.57 + \$756.14 + \$657.52 + \$571.75 + \$497.15$$

$$= \$3{,}352.16$$

This is what Uncle Joey's investment is worth to you right now, if you use a discount rate of 15%.

This kind of analysis, also known as a ***discounted cash flow***, or ***DCF***, is perfect for a spreadsheet. Make that computer do jumping jacks.

| | A | B | C | D |
|---|---|---|---|---|
| 1 | | | | |
| 2 | | Discount Rate | 15% | |
| 3 | | | | |
| 4 | Year | Cash Flow | Divide by | Discounted Cash Flow |
| 5 | 0 | 0 | 1.000 | $0 |
| 6 | 1 | $1,000 | 1.150 | $869.57 |
| 7 | 2 | $1,000 | 1.323 | $756.14 |
| 8 | 3 | $1,000 | 1.521 | $657.52 |
| 9 | 4 | $1,000 | 1.749 | $571.75 |
| 10 | 5 | $1,000 | 2.011 | $497.18 |
| 11 | | | | |
| 12 | | | NPV | $3,352.16 |

The "divide by" column is 1.15 raised to the relevant power: 1.15, $(1.15)^2$, $(1.15)^3$, etc.

Here's the same sheet with formulas shown.

| | A | B | C | D |
|---|---|---|---|---|
| 1 | | | | |
| 2 | | Discount Rate | 15% | |
| 3 | | | | |
| 4 | Year | Cash Flow | Divide by | Discounted Cash Flow |
| 5 | 0 | 0 | =(1+$C$2)^A5 | =B5/C5 |
| 6 | 1 | $1,000 | =(1+$C$2)^A6 | =B6/C6 |
| 7 | 2 | $1,000 | =(1+$C$2)^A7 | =B7/C7 |
| 8 | 3 | $1,000 | =(1+$C$2)^A8 | =B8/C8 |
| 9 | 4 | $1,000 | =(1+$C$2)^A9 | =B9/C9 |
| 10 | 5 | $1,000 | =(1+$C$2)^A10 | =B10/C10 |
| 11 | | | | |
| 12 | | | NPV | =SUM(D5:D10) |

After you type in the formulas in C5 and D5, you can just copy them down. They'll adjust on their own.

If you build this worksheet in Excel, try playing with the discount rate and see what happens to the NPV. As you lower the rate, the NPV goes up. If the discount rate drops all the way to 0%, then the NPV reaches the theoretical maximum of $5,000, because every future dollar is worth exactly $1 today. If you increase the discount rate, the NPV decreases, since future dollars are worth less in today's dollars.

## DEAL OR NO DEAL

Should you take Uncle Joey's deal? It all depends on the cost to buy in. You've already figured out that if you discount at 15%, all the future cash flows are worth $3,352.16 to you right now. Say you've compared this deal to other investments with various rates of return, and you're comfortable with applying a 15% discount rate in this case.

Then your decision is easy: if Uncle Joey offers you the deal for anything less than $3,352.16, you should jump on it. For instance, if he wants to charge you $3,000, then things look good from your point of view. You'll spend three grand right now and get $3,352.16 in present value. Your "instant" profit is $352.16, so you're overjoyed.

On the other hand, say Uncle Joey won't take a penny less than $4,000, because he says the deal is "practically a sure thing." Who cares what he says or thinks? *You* think that the project is risky enough (and that Uncle Joey himself is risky enough) to merit a 15% discount rate, and so you're losing money if you spend $4,000 to get $3,352.16 in present value. Leave the deal on the table at that price, even if he is your godfather.

People often calculate the NPV of an investment including not just the future cash flows but the initial outlay as well. In that case, represent that initial outlay as a negative cash flow.

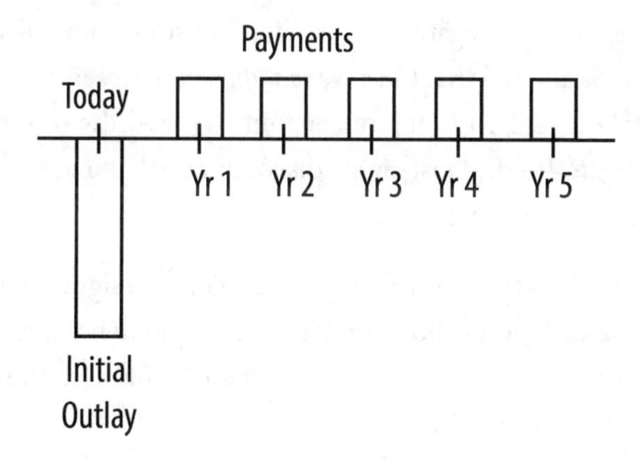

If you run the numbers with the initial outlay included, then the decision rule is simply this: take the deal if the NPV is greater than zero, and reject it if not. The NPV is essentially your instant profit on the deal, and you only want positive profits.

For instance, if Uncle Joey insists on a $4,000 up-front investment, then your NPV is $3,352.16 – $4,000 = –$647.84. Notice that the four grand is not discounted at all, because it's already in today's dollars. Since the NPV is negative, you reject the offer.

Your uncle sputters, "But I'm only asking for four grand, and you're going to get five back!"

Since you're taking finance, you use charts and graphs to explain to him the time value of money and net present value.

Uncle Joey scratches his prominent ears and says, "Hmm. If you choose a lower discount rate, the future cash flows will be worth more today, right? So let me ask you this. What rate would you have to use to make this deal worth your while?"

Your uncle has an interesting point. After all, if the discount rate were 0% (which is unrealistic, but just for giggles), then the $5,000 of future cash flows would be worth exactly $5,000 today, so you'd take the deal for a price of $4,000 and pocket a positive NPV of $1,000. Meanwhile, at 15%, the NPV is negative. Thus, there must be some discount rate between 15% and 0% that *leaves you indifferent* between taking the deal and not taking it. That indifference point is called the ***internal rate of return***, or ***IRR***. It's the "just-right" discount rate that makes the NPV of all the cash flows equal to zero.

You can use the HP 10b11 financial calculator to figure out the IRR. Input all the cash flows above, making sure to put in *negative* $4,000 for the first cash flow in Year 0, then click the IRR button. The calculator runs and spits out 7.93%.

This means that if you apply a discount rate between 0% and 7.93% to Uncle Joey's project, you'll get a positive NPV and take the deal. Conversely, if you apply a rate higher than 7.93%, as you did before, you'll get a negative NPV and pass on the deal.

Putting this another way, you compare the IRR of 7.93% to the discount rate you apply to similar investments. This comparison rate is called the ***hurdle rate*** or the ***opportunity cost of capital***. If the IRR is higher than the hurdle rate, then take the deal; otherwise, don't. For instance, you had decided earlier that the hurdle rate for this investment ought to be 15%. Since the IRR came in at only 7.93%, it didn't "clear the hurdle" of 15% that you had set, so you tell Uncle Joey no.

Incidentally, you can also figure out the IRR using the Excel spreadsheet from earlier. Just try different numbers in the discount rate cell till you get as close as you want to $4,000 for the NPV of just the future cash flows (since that spreadsheet didn't include the initial outlay). Or, as is described in Chapter 1, use the Goal Seek tool in Excel. On the Data ribbon, choose What-If Analysis, choose Goal Seek, then ask Excel to *set cell* D12 *to value* $4,000 *by changing cell* C2. *Voilà*! The answer is 7.93%, to a couple of decimal places.

For various technical reasons, the IRR rule isn't as good as the NPV rule ("take deals with positive NPV"), so stick to the latter. However, IRR is alive and well in industry, so you need to know how to find it and use it.

## ...JAMES BOND

Many bonds (the formal I.O.U.'s that companies and governments sell) have a periodic constant payment called a ***coupon***. These coupon payments end when the bond reaches ***maturity*** and dies. At the end of the bond's life, the ***principal*** or ***face value*** of the bond is typically paid back as well, whether or not the bond was actually purchased for that face value.

Take a $1,000 face-value bond with a 5% coupon paid every year for three years, at which point the bond matures. First, note that the 5% is *not* necessarily the discount rate you should apply! It's just a way of expressing the coupon as a percent of the face value of the bond:

$$\text{Coupon} = 5\% \times \$1{,}000 = \$50$$

You can lay out the cash flows you'd get, if you happened to own this bond.

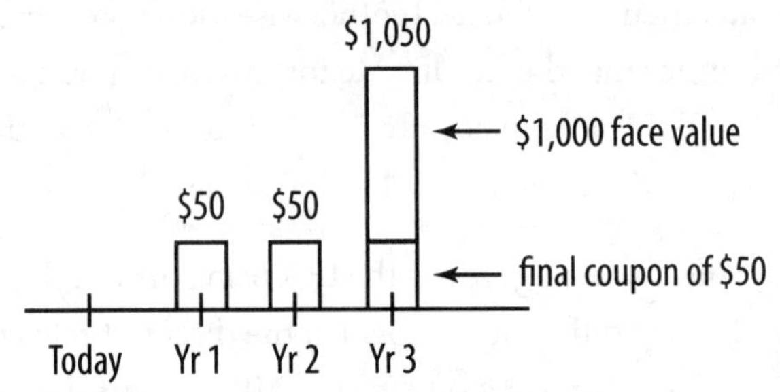


If you discount these cash flows at say 7%, you'll get an NPV of $947.51.

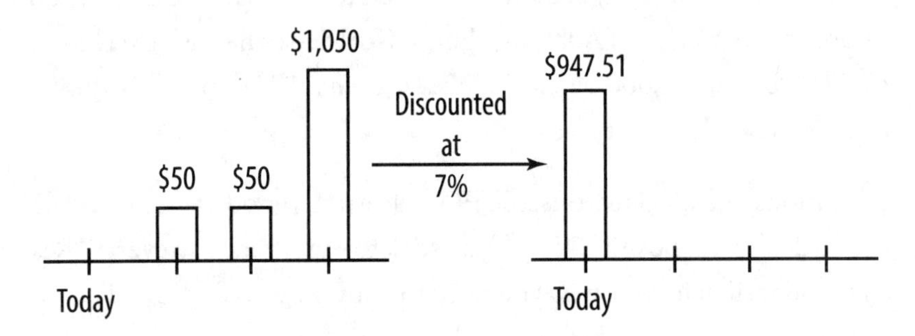


If you discount at only 3%, the NPV is $1,056.57.

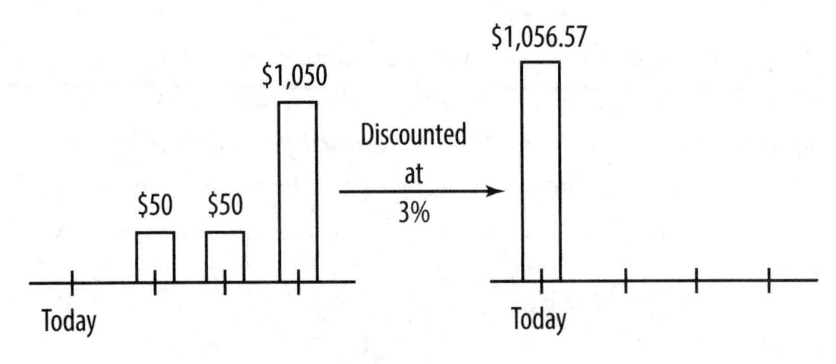

And if you discount at exactly 5%, which happens to be the coupon rate, then the NPV is exactly $1,000, the face value. Ta da!

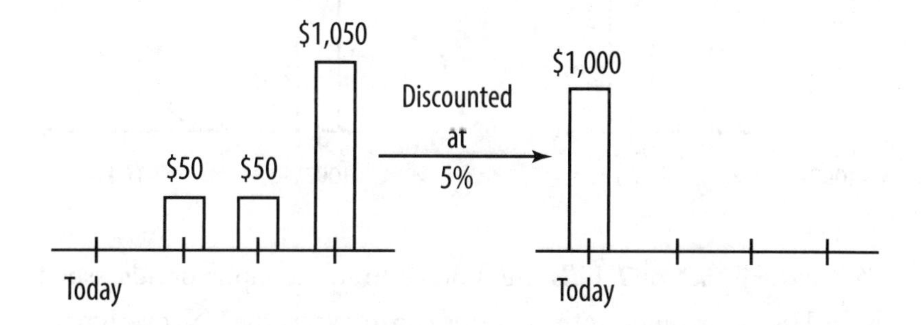


That's no accident. Something that pays you $50 a year like clockwork, then gives you an extra $1,000 at the end, is worth exactly $1,000 to you right now if the interest rate you're demanding is exactly 5%. Then the coupon is paying you the perfect amount of interest each year to keep you happy, and at the end you're also given your money back.

If, however, these cash flows are regarded in the marketplace (and by you) as riskier than comparable investments that pay 5%, then you and the other market participants will demand a discount bigger than 5%, and the current price of the bond falls below the face value (known as ***par***) of $1,000.

Likewise, if you and everyone else start to consider these cash flows *less* risky than comparable investments that pay 5%, the price will rise above par, and the yield of the bond (the applicable discount rate) will drop below 5%.

Finally, there are ***zero-coupon bonds*** that don't pay a coupon at all, but only return one payment at maturity. That payment could be the face value of the bond or the face value plus compounded interest. "Zeros" have a current price below the dollar amount of the final payment, unless something very weird's going on in the bond market.

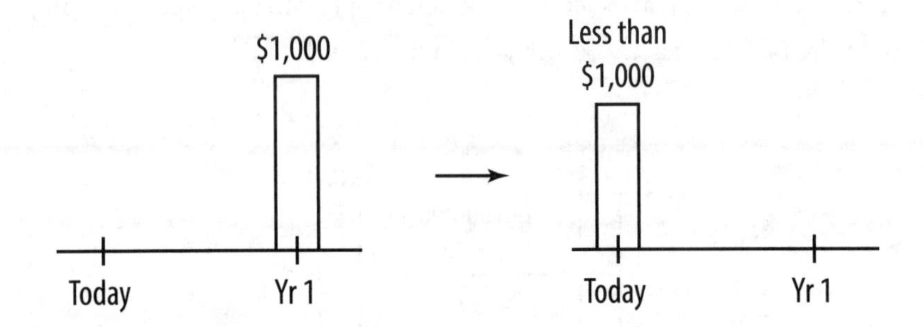


***US Treasury bills***, or ***T-bills***, are short-term zero-coupon bonds issued by the US government, with maturities up to a year. The US government also issues bonds with much longer maturities (up to 30 years). The yields on US government bonds with different maturities are usually not the same, a fact known as the ***term structure of interest rates***.

If you plot the interest rates of the various bonds, you get the ***yield curve***. Normally, this curve slopes upwards, because longer-term bonds typically pay higher interest than short-term bonds do. When the yield curve "inverts" or "flips," look for dragons flying out of the sky, because Armageddon (or at least a recession) is probably approaching.

## FOREVER'S A LONG TIME

It may seem strange, but you can measure the NPV of an *infinite* stream of payments:

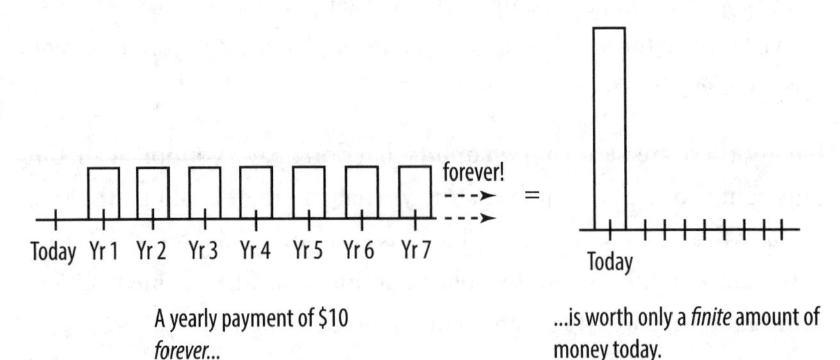

The reason this is possible is that the present value of payments in the distant future shrinks exponentially (in a very literal sense), as long as there's a positive discount rate. So this infinite set adds up to a finite number. It's just like saying that:

$$\frac{1}{2}+\frac{1}{4}+\frac{1}{8}+\frac{1}{16}+\frac{1}{32}+\overset{\text{forever!}}{\ldots}=1$$

If you go halfway across the room, then halfway again, then halfway again, and again, *forever*—you reach the other side.

Whoa. Deep.

Okay, so how do you do this infinite sum? Say you've got this promise of a yearly payment of $10 forever. You trust the issuer of this ***perpetuity*** a great deal, so you'll apply the relatively low discount rate of 5%.

You could set up an infinite spreadsheet, with an endless number of rows...

Discount Rate 5%

| Year | Cash Flow | Divide By | DCF |
|---|---|---|---|
| 0 | 0 | 1.000 | 0 |
| 1 | $10 | 1.050 | $9.52 |
| 2 | $10 | 1.103 | $9.07 |
| 3 | $10 | 1.158 | $8.64 |
| ⋮ | | | |

till kingdom come

...but that's not very practical. Or you could do a complicated mathematical derivation (cool but also impractical), or you could learn a simple formula.

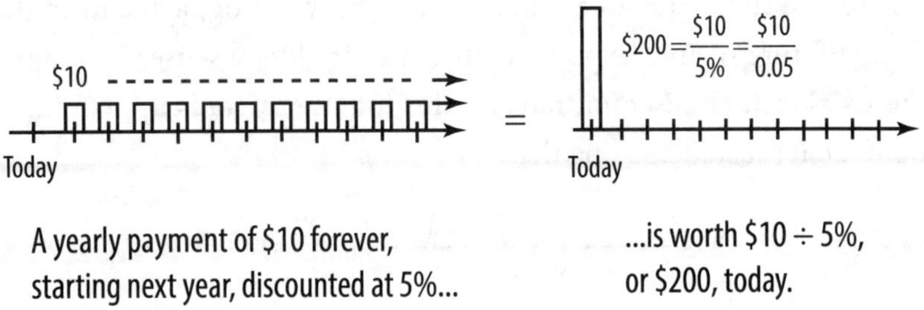

Amazing. That infinite set of discounted cash flows adds up to exactly $200. More generally, here's the formula:

$$\text{NPV of a perpetuity} = \frac{\text{Payment}}{\text{Discount Rate}}$$

$$\$200 = \frac{\$10}{0.05}$$

A little thought can reveal why this must be the case. This investment is going to pay you $10 a year indefinitely. If 5% is the discount rate you apply to this investment, then a $10 annual payment must represent 5% of the value of the investment.

$$\$10 = 5\% \text{ of the value}$$

$$\$10 = (0.05)(\text{Value})$$

$$\frac{\$10}{0.05} = \text{Value} = \$200$$

Notice how big a multiplier this is: 20 times the annual payment of $10 is the NPV ($200). That 20 is just 1 divided by the discount rate:

$$20 = \frac{1}{0.05} = \frac{1}{5\%}$$

The smaller the discount rate, the bigger the multiple, because future cash flows don't get discounted as much. So the whole stack is worth (relatively speaking) a lot today.

| Discount Rate | Multiplier | |
|---|---|---|
| $= 2\%$ | $= 1/0.02 = 50$ | A yearly payment of \$10 forever is worth $\$10 \times 50 = \$500$ today, if you discount at 2%. |

Conversely, the larger the discount rate, the smaller the multiple, because future cash flows get discounted a lot. So the whole stack is not worth very much today.

| Discount Rate | Multiplier | |
|---|---|---|
| $= 15\%$ | $= 1/0.15 = 6.667$ | A yearly payment of \$10 forever is worth $\$10 \times 6.667 = \$66.67$ today, if you discount at 15%. |

Notice *how sensitive* the NPV of a perpetuity is to the discount rate, especially when the discount rate is small. This is one way to fudge numbers to make them come out the way you want.

A key point to remember when you value a perpetuity is that the first payment always starts *next* year, in Year 1; today is Year 0. In fact, in any DCF (discounted cash flow) analysis, today is Time 0.

## A GROWING STREAM

What if that infinite stream of payments is itself getting larger over time? How much are they worth now?

Let's say that next year, you'll get a \$10 payment, and then every payment after that will be larger than the previous by 3%.

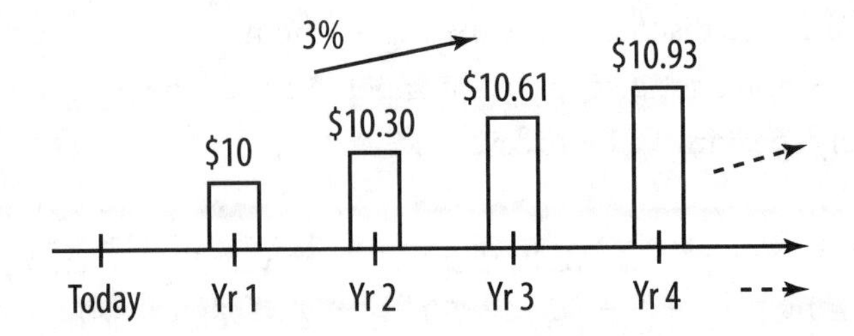


All you have to do is pick a discount rate that exceeds this growth rate. (To be conservative, pick small growth rates and large discount rates for perpetuities.) Say the discount rate $r = 8\%$. Then the math turns out to be easy.

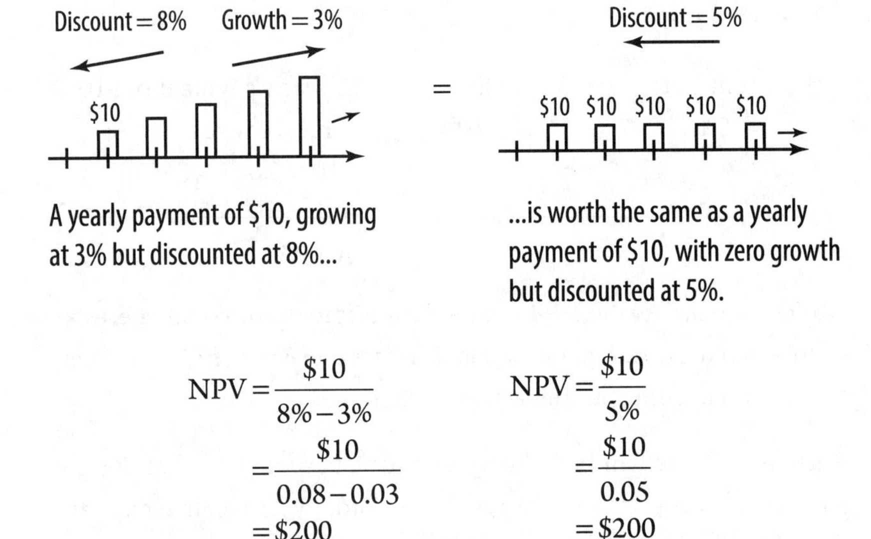


$$\text{NPV} = \frac{\$10}{8\% - 3\%} = \frac{\$10}{0.08 - 0.03} = \$200$$

$$\text{NPV} = \frac{\$10}{5\%} = \frac{\$10}{0.05} = \$200$$

In general, if the growth rate is $g$ and the discount rate is $r$, then:

$$\text{NPV of a growing perpetuity} = \frac{\text{Payment (next year)}}{r - g}$$

This is why you pick large $r$'s and small $g$'s. To avoid dividing by zero or a negative number, you must make $r$ bigger than $g$.

## STOCK VALUATION

Why all this focus on perpetuities? For one thing, because owning a share of stock in a dividend-paying company gives you the right to an endless stream of dividends (unless the company goes bankrupt, gets acquired, cuts its dividend, or whatever). And in theory, *the current value of the stock is the net present value of all expected future dividends,* discounted back at an appropriate rate—even if the company does not currently pay dividends! (If a company declares it will *never* pay dividends, then it's telling its owners they can *never* take any profits out directly. Bad news.)

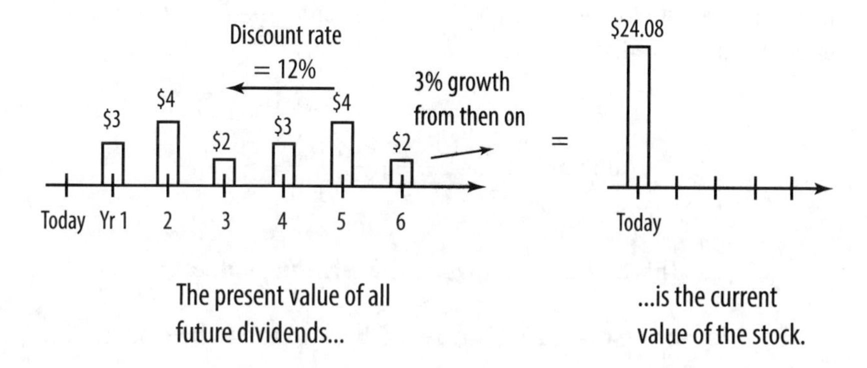


Here's how the process goes:

1. Estimate future dividends both in the near term and in the long term.

   At some point, you'll have to pick a final dividend and then estimate a continuing growth rate. Be careful—that final dividend ($2 in Year 6) and growth rate (3%) wind up contributing over half of the NPV of $24.

2. Compute the present value of all but the final dividend.

$$\text{PV of Years 1–5} = \frac{\$3}{1.12} + \frac{\$4}{(1.12)^2} + \frac{\$2}{(1.12)^3} + \frac{\$3}{(1.12)^4} + \frac{\$4}{(1.12)^5}$$
$$= \$11.47$$

3. Compute the present value of Year 6 and later dividends in two steps:

   a. Bring the growing perpetuity back to a Year 5 value. Remember that perpetuities are valued using "next year's" payment as the first payment, so if Year 6 is "next year" for the moment, then Year 5 is "this year."

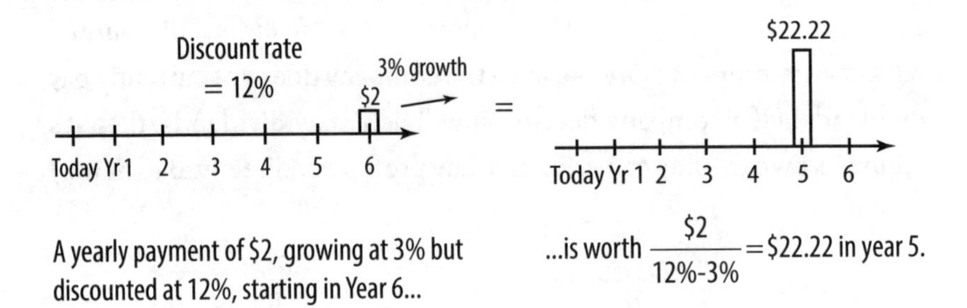


$$\text{Value in Year 5} = \frac{\text{Payment in Year 6}}{r-g}$$

This $22.22 is known as the ***terminal value***.

   b. Now bring that $22.22 in Year 5 back to today, by discounting at 12% over 5 years.

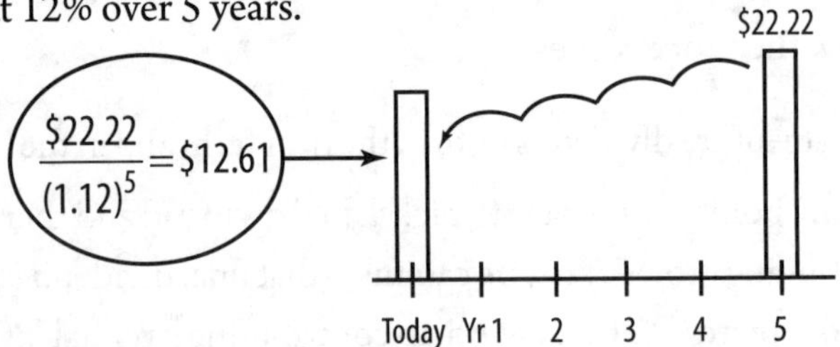


4. Add up the two large present values:

| $11.47 | + | $12.61 | = | $24.08 |
|---|---|---|---|---|
| PV of Years 1–5 | | PV of Years 6–Infinity | | |

Alternatively, you might use something called ***free cash flow*** (***FCF***) to value a company.

| Free cash flow (FCF) | = | Cash flow from operations (money thrown off by the operating activities of the company) | – | Capital expenditures (big investments by the company in equipment, etc.) |
|---|---|---|---|---|

Free cash flow is the cash available to pay back investors (both debtholders and shareholders). The current value of a company, including both debt and equity, is the net present value of all predicted FCF's. It is hard to game these FCF numbers on accounting statements, and they provide better insight into companies that don't pay dividends currently or have never done so. You do the same kind of DCF analysis on free cash flows as you do on dividends.

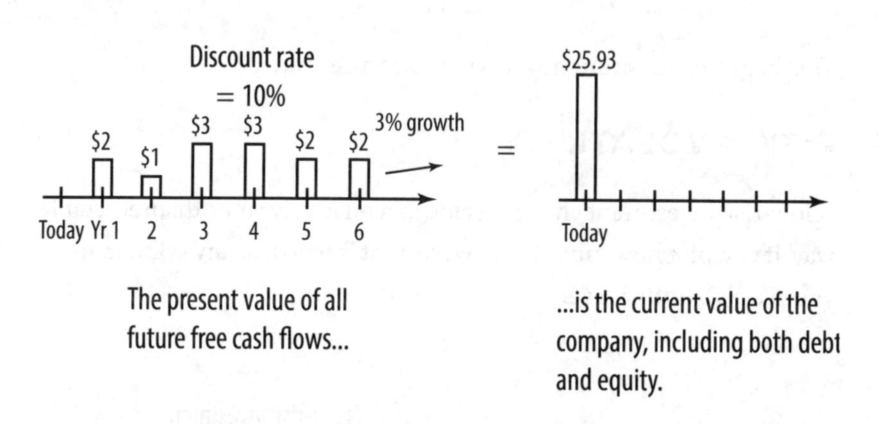


There are other ways to value stocks besides fundamental analysis of cash flows. Most importantly, you can compare the company to others in its industry. For instance, you might simply look at the ***P/E ratios*** (the ratio of stock price to earnings per share) of a number of competitors. Then you can determine whether Company X is expensive or not. The P/E tells you how many dollars you need to spend right now to buy a dollar of earnings, so a higher P/E means more expensive earnings.

When you think about this kind of ratio, however, don't abandon your knowledge of discounted cash flows. The P/E ratio works like a multiplier on a perpetuity. That is, a P/E of 20 times earnings means that profits represent only 5% (= 1/20) of its stock price. A higher P/E means a lower percent return in this sense.

## WHAT DISCOUNT RATE SHOULD YOU USE?

The key principle in picking an *r* for a set of future cash flows is this:

> Consider those cash flows as promises—as an investment you can buy into. Apply the same discount rate as you would to similar investments with similar risks.

Go back to the idea of risk and return. Riskier investments need to promise a higher expected rate of return, in order to attract money. Lower-risk investments do not need to pay as much.

This begs the question: how do you measure risk?

## RISK = VOLATILITY

A low-risk investment changes value in a relatively smooth, predictable way. If you plot how much that investment is worth at any point in time, you see a smooth curve:

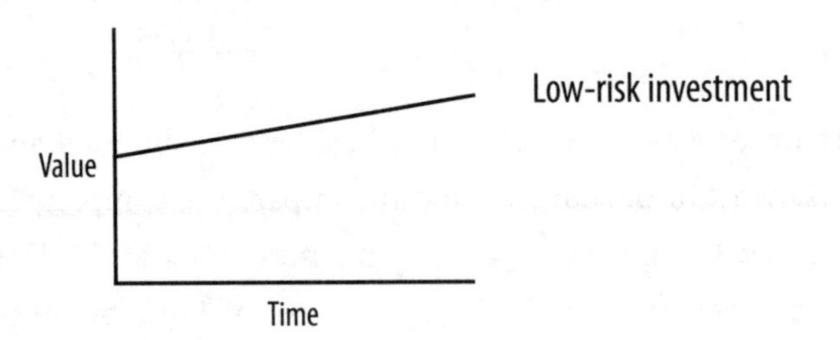


In contrast, a high-risk investment changes value in a jagged, unpredictable way:

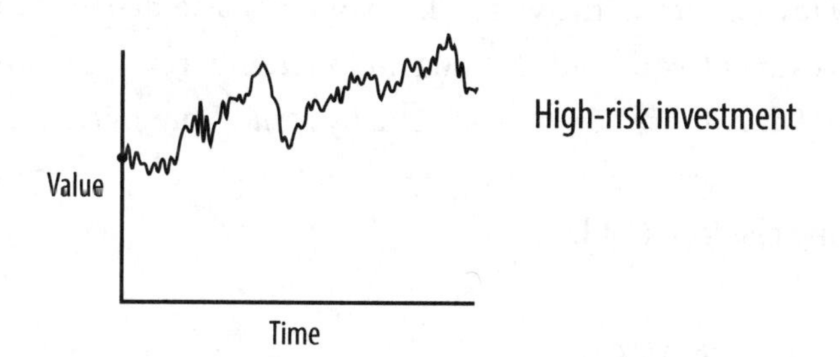


You can quantify this contrast. Gather up all the daily ***returns*** on each investment—the daily percent changes in the value of each.

| Date | Daily Return<br>Low-Risk | High-Risk |
|---|---|---|
| March 1 | +0.07% | +0.7% |
| March 2 | +0.07% | +0.02% |
| March 3 | +0.08% | −0.87% |
| March 4 | +0.07% | +1.28% |
| March 5 | +0.06% | −0.56% |

The low-risk returns are very steady, with low volatility, while the high-risk returns are all over the map—they're highly volatile.

Statistics rides to the rescue. As discussed in Chapter 3, the ***standard deviation*** of each set of returns provides a convenient measure of volatility over the relevant time period.

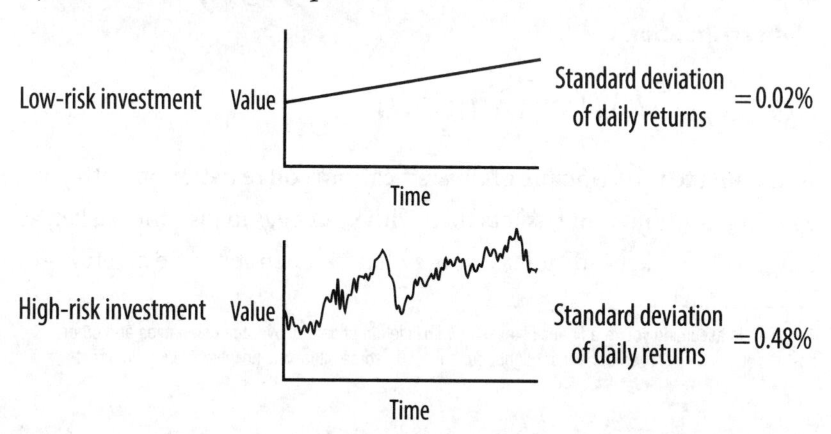

However, this approach doesn't go far enough. Standard deviation on its own just tells you about the risk of this one investment, as if it were the only investment in the world. But your risk also depends on every *other* investment you hold. Adding an inherently risky investment to your ***portfolio*** of investments can actually *reduce* your overall risk. This happens most clearly if this new investment tends to move opposite to everything else you hold:

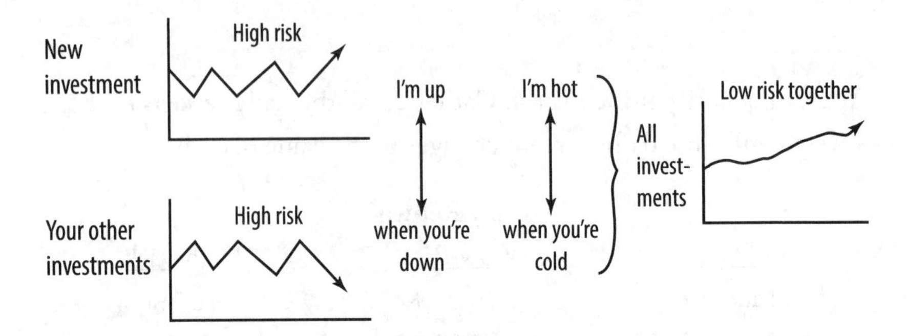


Many songs about toxic love relationships, such as "Hot 'N' Cold" by Katy Perry or "Love on the Brain" by Rihanna, are really about ***negative correlation***. A new investment that's negatively correlated with your other investments is a great thing, because you can raise your expected return while lowering your overall risk. This is the basic principle behind hedge funds, which theoretically try to hedge their bets.

You don't have to have negative correlation to win. A group of ***uncorrelated*** investments that move randomly relative to each other, or even investments with imperfect positive correlation, give you the benefits of ***diversification***.

## EGGS IN MANY BASKETS

According to the principle of diversification, you're better off with your eggs in many different baskets than with your eggs in just one basket, as long as the baskets are at least somewhat independent.[1] Be careful—in

1 This is assuming you're a financial investor with little direct control over the companies and other assets you invest in. A game theorist might counsel you in some situations (in which you have direct

times of market stress, previously uncorrelated values can become very correlated. No one thought that real estate prices in 20 different US cities would all decline at once—but they did. Lots of baskets got into one big handbasket and then all went to hell together.

## THE CAPITAL ASSET PRICING MODEL, OR CAPM

Back to the big question—what discount rate should you use to evaluate some potential investment, say the stock of some company?

The CAPM (pronounced "Cap-Em") gives you a process. To calculate the right discount rate for the stock, you need three pieces of information:

1. The ***"risk-free" rate of return,*** $r_f$

   This is the return you could get from a theoretically riskless investment. Sadly, there is no such riskless thing in reality. Short-term US government debt (Treasury bills) is taken as the proxy. (If the US government ever defaults on T-bills, get ready for a wild ride, Mad Max style.) Depending on your time frame, you might take $r_f = 5\%$ to 7%. Recently, the risk-free rate has been much lower.

2. The ***market rate of return,*** $r_m$

   This is what you'd earn by holding a basket of claims on every possible asset in the world. Sadly, a complete ***market basket*** also doesn't exist. The return on the ***S&P 500,*** a basket[2] of common stocks of the 500 largest publicly held US companies, is often taken as the proxy. Alternatively, you can take the ***market risk premium*** (conveniently defined as $r_m - r_f$), which historically has been around 8%, and add it to your value for $r_f$ to get $r_m$.

control) to put all your eggs in one basket and guard that basket with your life. This is known as *strategic commitment.* Whatever you do, don't count your eggs before they hatch.

2 It is a mystery why finance folks like baskets, whereas consultants prefer buckets.

3. The ***beta*** (Greek letter $\beta$) of the stock

   Beta is a number that answers this question:

   If the whole stock market twitches upwards 1%, what do you expect this particular stock to do, on average?

   High-beta stocks tend to twitch upward more than 1%—say 2% or even 3%. Tech stocks typically have high betas.

   Low-beta stocks don't tend to twitch upward as much as 1%. Maybe they'd tend to go up only 0.5%, or not move at all (zero beta), or even go negative, although that's really rare. Utilities have low betas.

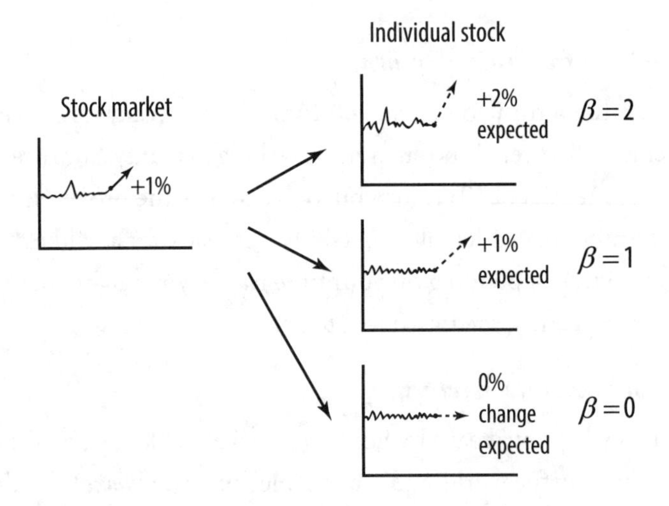


Like standard deviation, beta measures risk, but in a different way. Standard deviation measures the risk of an investment in isolation. Beta tells you what happens to your overall risk when you add some new investment to your portfolio. Beta is a measure of ***nondiversifiable risk***—the risk you can't simply get rid of by holding a bunch of different assets. By the way, you're generally presumed to be holding a portfolio consisting of some blend of the whole market basket and the risk-free asset. In theory, that's always supposed to be the best you can do.[3]

Anyway, the beta of stock X is itself the product of two numbers:

1. The correlation between the price of stock X and the overall level of the market.
2. The relative volatility of stock X, compared to the average volatility of the market. This is expressed as a ratio of standard deviations.

A high beta can come from a high correlation between the individual stock and the market (they tend to move together).

A highly volatile asset, though, could be completely uncorrelated with the overall market. In that case, the zero correlation would cause the asset to have a zero beta.

Once you have a beta for the stock (sometimes you just look it up or average the betas of comparable companies), you make a graph, with beta along the *x*-axis and return along the *y*-axis.

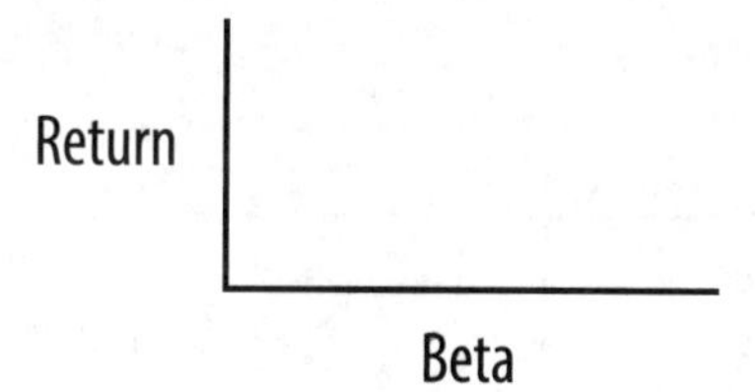


You put two points on the graph. One is for the risk-free investment. Since the risk-free asset has no risk, its volatility is zero, and its beta is also zero. However, it has some positive return, the risk-free rate $r_f$

3 Warren Buffett, Steve Cohen, and other billionaire investors would say otherwise, of course. Buffett and Cohen follow very different investing philosophies, but there's nothing odd about what they primarily invest in—the common stock of US companies—and they both have beaten the market handily, year after year.

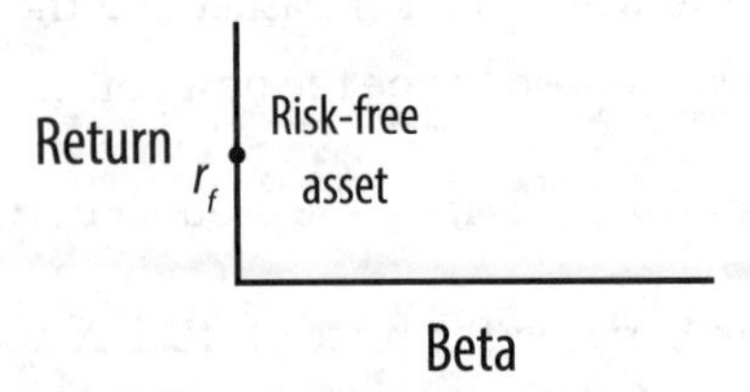


The other point is for the market basket. This asset has a return of $r_m$ and a beta of 1, by definition. (Beta measures *correlation with* and *volatility relative to* the market. The market is always perfectly correlated with itself and is as volatile as itself, so the market beta = 1.)

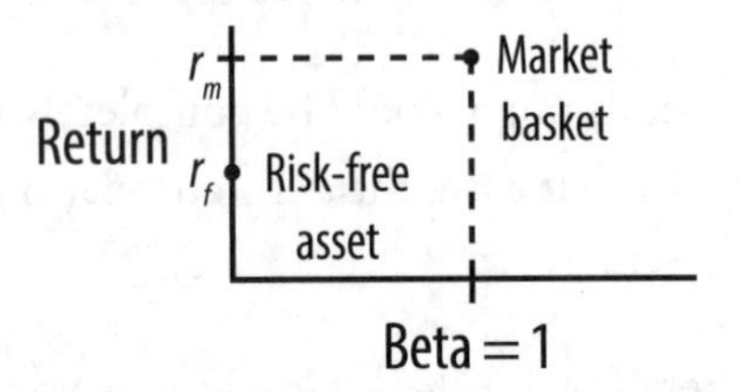


Draw a line between the two points. The expected return of any stock—what we've been looking for—lies on that ***security market line***. You tell us the beta (on the *x*-axis), and we'll tell you the expected return (on the *y*-axis).

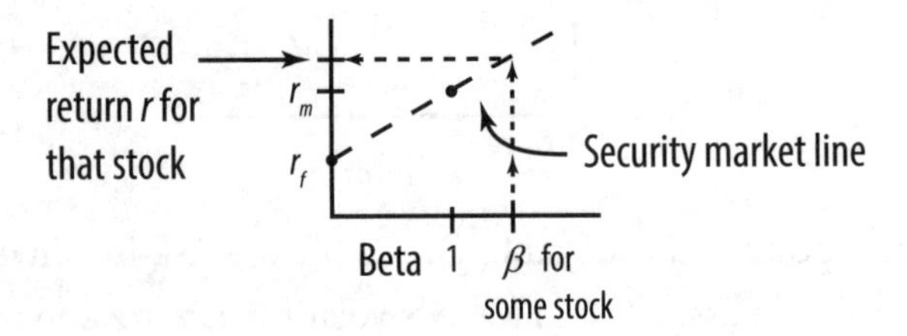


This is the heart of CAPM. As a formula, the security market line looks like the formula for a line.

| $y$ | $=$ | $mx$ | $+$ | $b$ |
|---|---|---|---|---|
| vertical axis | | slope ↗ ↖ horizontal axis | | $y$-intercept |

$$r = (r_m - r_f)\beta + r_f$$

Conceptually, CAPM is a fancy way of expressing the typical risk-return trade-off:

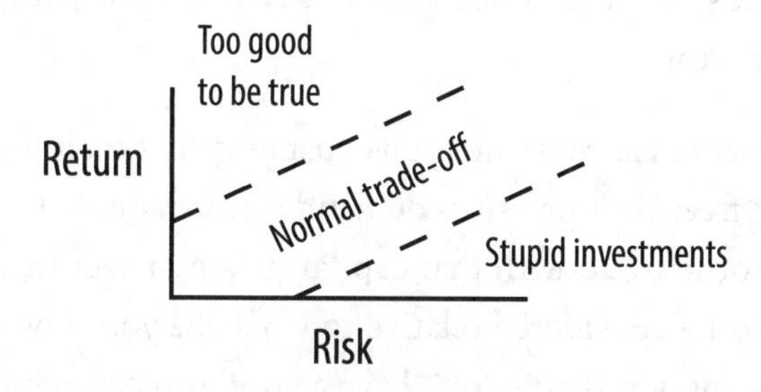


## WHAT ELSE IS THERE IN FIRST-YEAR FINANCE?

### 1. *Efficient market hypothesis (EMH)*

This says that big capital markets, such as that for ***large-cap US equities*** (the common stock of big US companies), are very competitive places. Everyone else in the game is probably as smart as you are. So if you've spotted what you think is a mispricing on some stock, the EMH would say you're wrong—***the price is right,*** as Bob Barker might say, because it already incorporates all the information and insights anyone could have.

The battle is on between efficient-market theorists and proponents of the new field of behavioral finance, which seeks to account for market bubbles and panics rather than command them to hide or not to exist in the first place. You won't do more than give a head nod to the existence of behavioral finance in an introductory course, though, so don't get all excited.

2. ***Capital structure***

Is your company funded more by debt (liabilities) or equity (stock ownership)? This is the question of capital structure. In a perfect world, the ***MM*** theorists (Miller & Modigliani, who never had a tiger show in Vegas) would argue that it doesn't matter. In the real world, or at least in the US, debt financing gets a tax break—companies don't get taxed on interest payments, but profits due to equity shareholders are taxed first at the corporate level. So the ***weighted average cost of capital***, or WACC, blends the cost of debt and the cost of equity together with that debt tax break.

In general, finance class reminds you constantly that capital—money to invest—isn't free. Your investors demand a return on that capital. And what you choose to do with that capital, as a manager in a company, should always be considered relative to what *else* your investors could do with their money. If you don't have a good project to invest in, one that has a positive NPV (meaning it beats out industry-standard projects), you should give your investors their money back, or sleep with one eye open.

3. ***Options***

A financial ***option*** is the right to buy or sell some asset at a certain price. For instance, a ***call option*** on stock X with a ***strike price*** of $10 is the right to *buy* the stock at $10, either at some future date or up until then. (A ***put option*** is the right to *sell* the stock. Think *call*—you can *call* it to you when you want. *Put*—you *put* it to someone else when you want.) Options are side bets on the direction you expect a stock price to go. Two key points to remember about options:

   a. ***Put-call parity***: The prices of calls and puts are linked together with the price of the underlying stock and the risk-free rate at which you can borrow money. If you know three of these numbers (price of a call, price of a put, price of the stock, and the risk-free rate), you can figure out the fourth.

b. ***Volatility***: The higher the volatility of the underlying stock, the more valuable the option. That may seem counterintuitive, until you realize that holding an option has *no downside* (except for the cost of the option in the first place). If you've got a call option with a strike of $10, you're betting the price of the stock will go up (that way, when you *exercise* the option and buy the stock at $10, you can resell it right away for more). So if the price goes up, that's what you do, and you make money. But if the price goes down, so what? You let the option expire unexercised. So you *like* volatility in the stock, because you don't care that that means more potential downside for equity holders (you're not one of them). You've taken a one-sided bet, and the more extreme the possible upside, the better for you.

# Chapter 6: Marketing

This chapter is deliberately short, since the typical first-year course doesn't contain too much quantitative content (the focus of this academic section). However, it's worth introducing a few key concepts and tools.

## WHAT IS MARKETING?

Marketing likes to consider itself distinct from sales, which has a hard, pushy connotation. Marketing is all about ***customer needs***—figuring out those needs, whether they're explicit or not, and then meeting those needs in a profitable way. In other words, the ***value*** of your product to your customer must be higher than its price to him or her. Meanwhile, the sale price has to more than cover *your* costs so that you can make a profit.

As a subject of study and as a profession, marketing is not soft and fuzzy. It's *very* quantitative and data driven. You might think that marketers are "creatives" dreaming up funny new ad campaigns. Yes, marketers need to be creative, but they are squarely on the business side of things—and no, it's not their job to come up with funny ads. That's what Mad Men are for.

## WHAT DO MARKETERS DO?

### *STP: Segment, Target, Position*

Marketers cut up large groups of customers into smaller, more homogeneous ***segments***. Then they ***target*** certain higher-priority segments and ***position*** products to appeal to those segments.

### *Marketing Mix: The Four P's*

The way marketers position the company's offerings is by adjusting elements of the ***marketing mix***, the levers available to marketers. These levers are commonly labeled the Four P's[4]:

1. ***Product***: its actual features
2. ***Price***: not only the direct price but also discounts, etc.
3. ***Place***: where the product is available, or the ***channels*** from you to the customer
4. ***Promotion***: advertising, public relations, etc.

### *Both Rational And Nonrational Appeals*

Good marketing works on every level. It takes into proper account the rational drives of consumer behavior—for instance, the fact that price cuts typically lead to increased demand. However, marketing also seeks to understand how real consumers are not perfect "utility maximizers" and to capitalize on that understanding. In this regard, marketing has a lot in common with social psychology and behavioral economics. Ultimately, whether by rational or by nonrational means, marketers want customers to "sign on the line that is dotted," as Alec Baldwin's character proclaims in *Glengarry Glen Ross*, an astounding film and play about real estate salesmen.

4 Developed by Jerome McCarthy. So simple, so brilliant. Our competing model "The 13 X's" has not caught on as well.

*Glengarry* also provides two marketing-related acronyms:

| | |
|---|---|
| A | Always |
| B | Be |
| C | Closing |

ABC expresses what you always gotta be doing.

| | |
|---|---|
| A | Attention |
| I | Interest |
| D | Decision |
| A | Action |

The AIDA model was actually published in 1925 by E. K. Strong, although the D stood for Desire. Either way, AIDA is actually a useful little mnemonic for the stages of the buying process: first, you need the customer's *attention,* then you pique his or her *interest* and awaken *desire,* until the customer makes the *decision* and takes *action.* Of course, Alec Baldwin delivers this with more oomph.

## DEFINING AND SIZING A MARKET

The word ***market*** has different meanings in different fields.

- In finance, it's never "a" market—it's always "the" market or "the" markets, meaning the financial markets, where financial securities are bought and sold. In the same way, to Wall Streeters, "the city" only means one city: New York (unless you're in London, in which case it's The City).
- In economics, a market is a collection of both buyers and sellers, buying and selling something.
- In marketing, the term ***market*** usually refers just to the buyers—the ***current or potential customers of a product*** or set of products.

This immediately brings up a question of scope:

- How broad is the geographic range you care about?
- How broad is the product definition?

| Narrow → | Broad |
|---|---|
| The market for children's DVDs in Atlanta | The market for children's entertainment in the Southeast |

Once you've decided on a geographic and product scope, you can ***size*** the market:

- How many customers are there or could there be? (Less common)
- How much money do those customers spend annually or could they spend annually? (More common)

So you'll hear these kinds of remarks:

"The market for children's DVDs in Atlanta is $X million."

How do you get to $X million? By multiplication:

$$\text{Market for children's DVDs in Atlanta} = \left(\text{Number of customers in Atlanta}\right) \times \left(\text{Annual spending on children's DVDs per customer}\right)$$

You can break up the factors on the right further. ***Households (HH)*** are often a convenient "customer" to think about.

$$\left(\text{Number of customers in Atlanta}\right) = \left(\text{Population of Atlanta}\right) \times \left(\frac{1 \text{ Household}}{\# \text{ of People}}\right) \times \left(\%\text{ of HH with young children}\right) \times \left(\%\text{ of those HH that buy}\right)$$

Each of these terms on the right of the equation can be either looked up or estimated.

Try an estimation:

$$\left(\text{3 million people in Atlanta}\right) \times \left(\frac{\text{1 Household}}{\text{3 People}}\right) \times \left(25\%\right) \times \left(80\%\right)$$

- 3 million people in Atlanta: Making this up
- 1 Household / 3 People: Also making this up
- 25%: Assuming children between 2 and 15, about 1/4 of the population
- 80%: High (but not 100%) ***penetration*** of DVD players

$$= \text{200,000 households in the market}$$

Now break up the annual spending per household:

$$\text{Annual spending on children's DVDs per customer} = \left(\text{Number of purchases per year}\right) \times \left(\text{\$ per purchase}\right)$$

$$= \left(\text{10 DVDs}\right) \times \left(\$15\right)$$

- 10 DVDs: Maybe estimated too high
- $15: Average retail, estimated

$$= \text{\$150 per household}$$

So this is what you get from this back-of-the-envelope calculation:

$$\left(\text{Market for children's DVDs in Atlanta}\right) = \left(\text{Number of customers in Atlanta}\right) \times \left(\text{Annual spending on children's DVDs per customer}\right)$$

$$= \left(\text{200,000 HH}\right) \times \left(\$150/\text{HH}\right)$$

$$= \text{\$30 million}$$

If you want to do that math in your head, borrow a 0 from the $150 to make 2,000,000 on the left, then multiply:

200,000 × $150

2,000,000 × $15

2 million × $15

= $30 million

If your ***market share*** is 10%, then you're selling $3 million of the children's DVDs to the Atlanta market. In real life, you do all these estimates and calculations much more precisely. But you have to be able to scribble something on the back of any envelope.

## QUANT IN MARKETING

Although the typical first marketing course isn't that quant-heavy, a number of quantitative tools are used by working marketers. You could say that marketing is an arena in which many quant tools born in other subjects come out to play.

1. ***Economics and marketing***

Marketers often try to figure out the actual ***demand curve*** for their products and/or for the industry. One way to do this is with little experiments with discounts. The information about price elasticity of demand is well worth the cost.

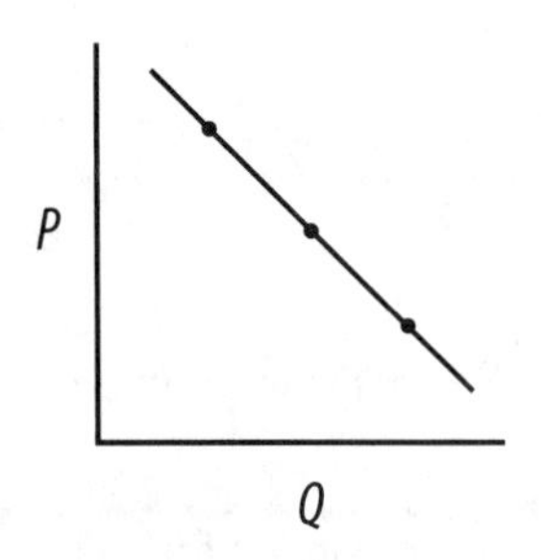


Marketers may also try to find the customer ***utility function***, which measures the typical customer's utility or happiness as a factor of various product attributes. In other words, you have a mathematical representation of customer preferences.

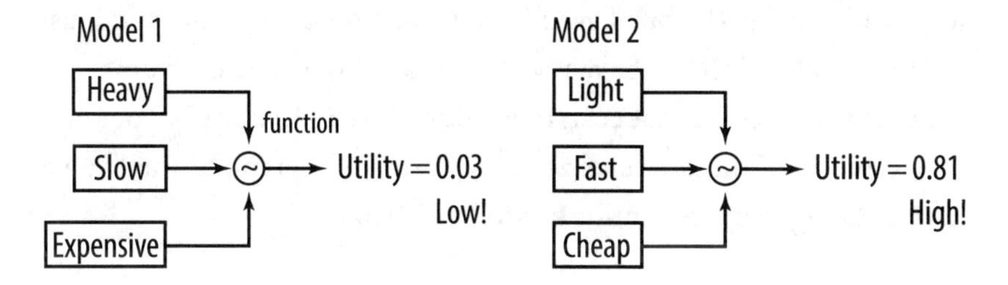


Model 1 is obviously worse (= has lower utility) than Model 2; you don't need a utility function to know that. But what about a heavy, cheap model vs. a light, expensive model? (Think laptops.) Now the utility function gets interesting. Using such a function to represent consumer preferences can shed light on tough decisions, such as trading off weight and cost.

Finally, a quant concept frequently used in economics is that of ***marginal benefit*** and ***marginal cost***. In marketing, theoretically you'd want to optimize your marketing mix by equalizing the marginal benefit of the last dollar you spent on each part of the mix:

| Marginal Benefit (MB) of final $1 spent to improve Product attributes | = | MB of final $1 spent on Pricing elements (e.g., discounts) | = | MB of final $1 spent on Place elements (e.g., channel incentives) | = | MB of final $1 spent on Promotion (e.g., advertising) |
|---|---|---|---|---|---|---|

It might be difficult to measure all these marginal benefits, but the thinking exercise is worthwhile anyway.

## 2. *Accounting, finance, and marketing*

To measure progress and success, marketers use metrics defined in accounting and finance, such as ***contribution margin*** (price minus variable cost) and ***return on investment*** (profit divided by investment, to simplify matters radically). For instance, you might use the contribution margin and a product's fixed costs (FC) to do a quick ***breakeven analysis***. Say that you can sell the product for $100, while the variable cost is $60. Then the contribution margin is $100 – $60 = $40, which is the amount each sale contributes to cover fixed costs and ultimately to provide profit. The breakeven (BE) is the number of units you must sell to cover the fixed cost exactly. If FC = $400,000, and each unit contributes $40 to cover that cost, then your BE is $400,000/$40 per unit = 10,000 units. You can visualize this situation on a graph, as was shown in the Managerial Accounting section of Chapter 4:

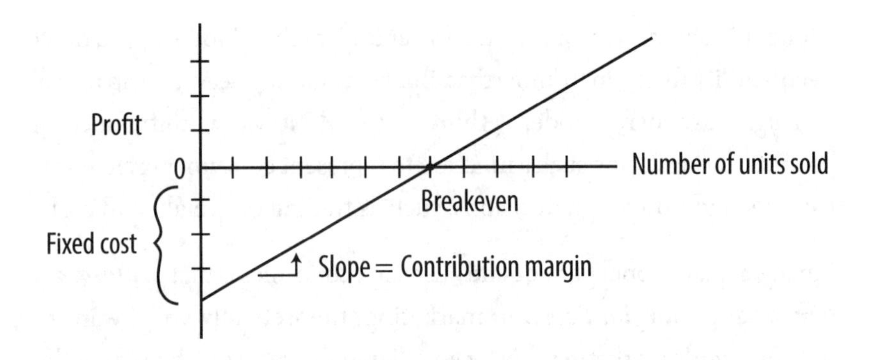


Accounting is the language of business, they say, and marketers must speak that language. Finance provides a means of evaluating possible future cash flows from projects, such as a potential new product launch. For instance, if your analysis of this launch shows a positive ***net present value***, after you discount future cash flows at an appropriate rate, then finance says you should pull the trigger and launch the product. As you monitor the launch, you would study ***sales variance*** reports to examine how reality matched your forecasts.

3. ***Statistics and marketing***

Statistical tools are of extreme importance in marketing. The first tool shows up in first-year stats class:

a. ***Regression analysis*** permits you to forecast demand and predict the sales response to various marketing levers you might pull, such as increased spending on promotions.

The next few tools probably won't be covered in your stats course, but you should understand what they are used to do.

b. ***Factor analysis*** (or principal components analysis) reduces a very complex picture of your customers to a simpler one. Say you know 50 separate pieces of information about each of your customers. That's a 50-dimensional space—way too hard to conceptualize. Factor analysis can help you reduce the number of variables, maybe down to three or four. Moreover, those three or four variables might have meaningful interpretations themselves, shedding light on your customer base.

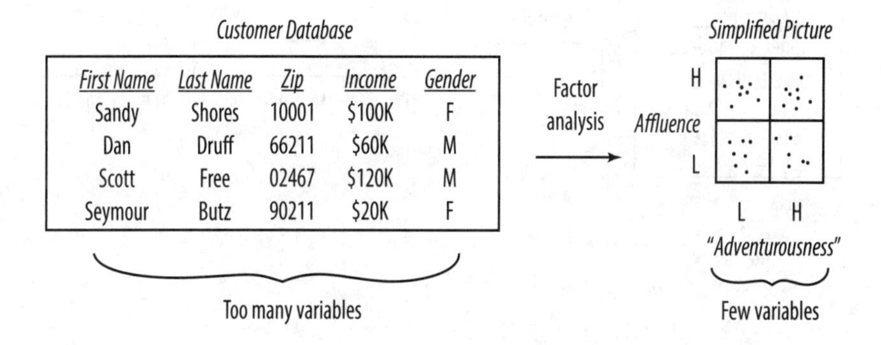


c. ***Cluster analysis*** does what it says—it helps you separate data into natural clusters.

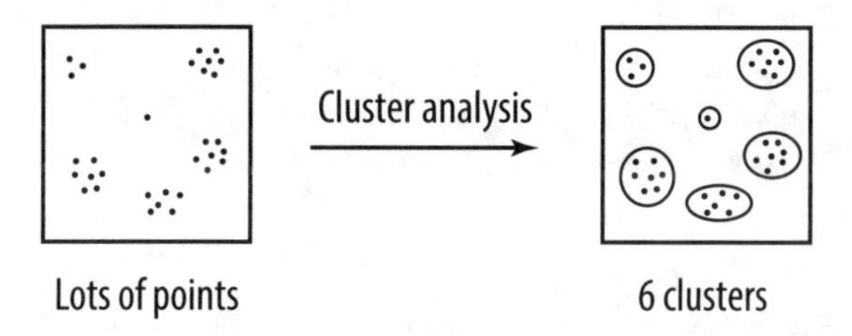


These clusters might correspond to market segments. Customers within a segment are similar to each other but different from customers in other segments. Clusters can provide the appropriate grouping.

d. ***Conjoint analysis*** lets you efficiently uncover customer preferences, as expressed through utility functions. You ask customers to pick between various models of a product. Those models embody different possible attributes, such as price, speed, color, etc. The nice thing is that you don't have to ask about every possible combination of attributes (which would take way too long). You can discover customer preferences with just a few well-chosen questions.

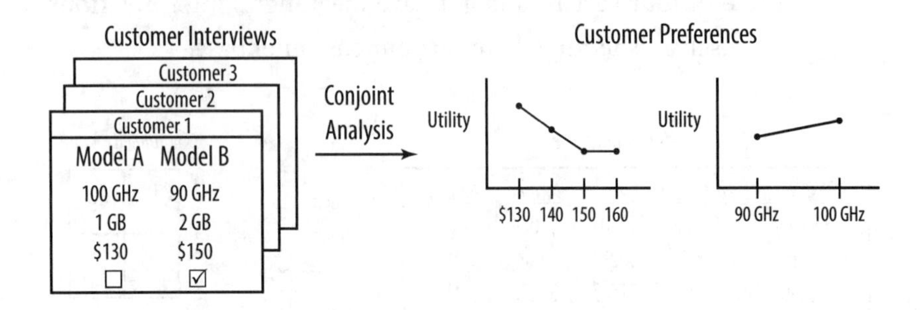

# Chapter 7: Operations & Supply Chain

If you like the "business = factory" metaphor that we discussed earlier, you've come to the right place. Operations and supply chain management are all about making the factory sing with efficiency.

"Ops" is a very broad subject, but this chapter focuses on three core quantitative topics:

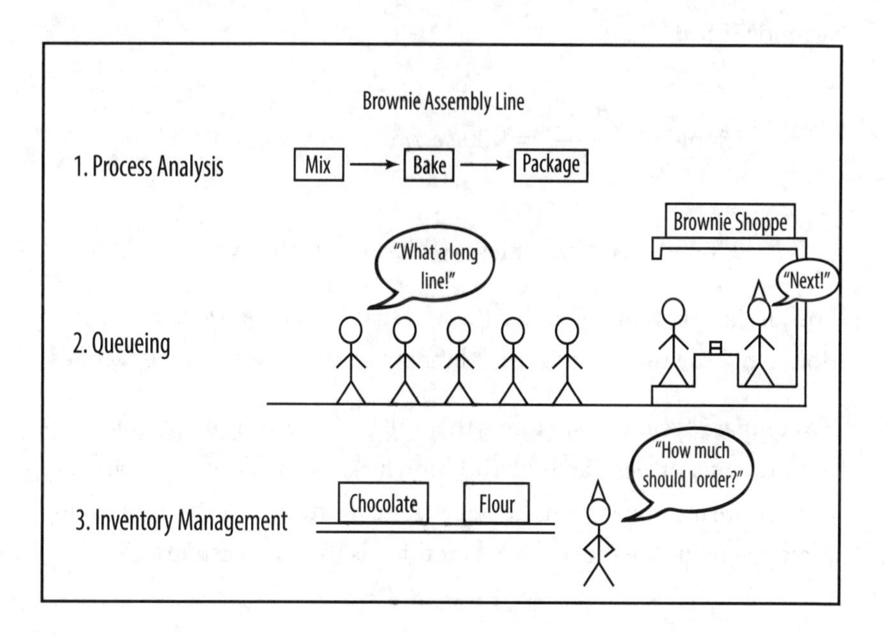

## PROCESS ANALYSIS

Processes are recipes or assembly lines with stages. Say you're making a lot of brownies to sell.

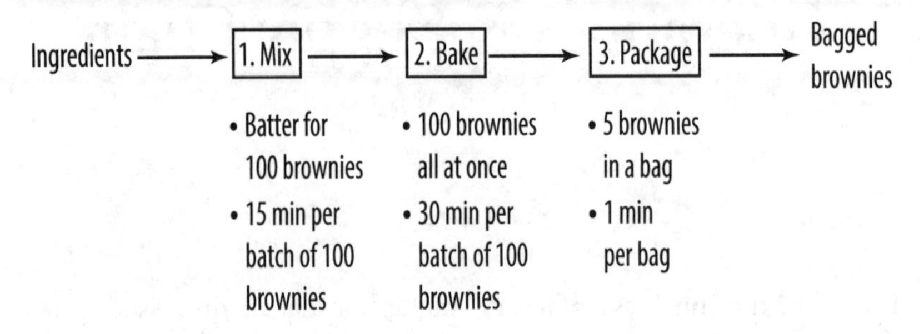


Lay out what happens at each stage and how much time it takes—the ***cycle time***, or ***CT***, for that stage. For instance, the cycle time for the Mix is 15 minutes per batch of 100 brownies. You can leave it like that, or you can convert to a "per brownie" time in either minutes (min) or seconds (sec):

$$15 \text{ min} \times \frac{60 \text{ sec}}{\text{min}} = 900 \text{ seconds}$$

$$900 \text{ sec} \div 100 \text{ brownies} = 9 \text{ seconds per brownie}$$

You just have to remember that the Mix stage is a ***batch process***—a single dollop of brownie batter doesn't literally come out every nine seconds.

To compare cycle times, express them all with the same denominator—per brownie, or per batch of 100 brownies. Just pick one. Batches are often more intuitive when the stages actually involve them (as in baking examples), but the most important thing is to be consistent.

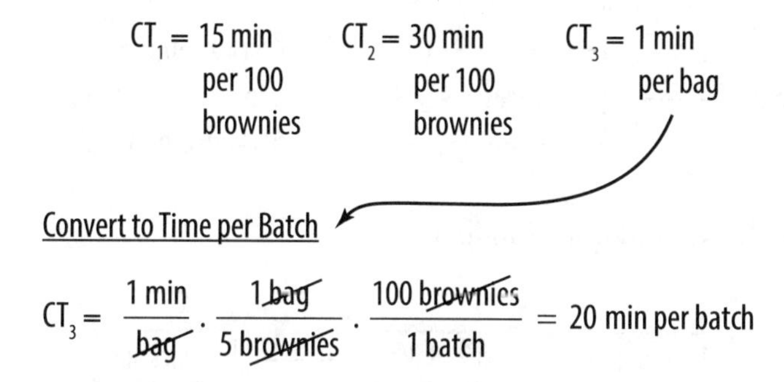


How fast can this assembly line crank out a batch? How much time does one batch of brownies take?

First, you have to clarify what you mean, since there are two possibilities.

1. Is the question "How long does a particular batch take *from start to finish*?"

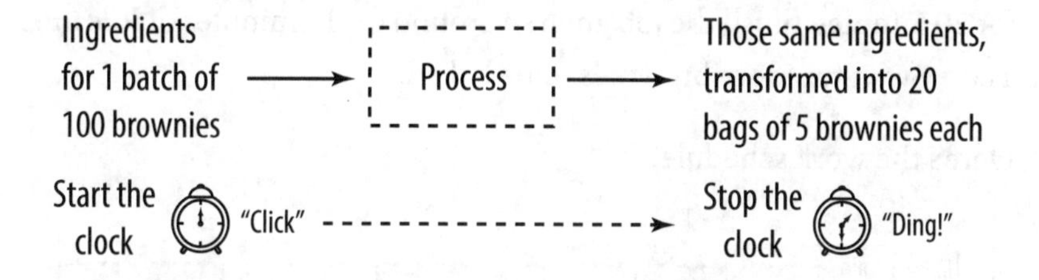


This is called the ***manufacturer's lead time, MLT***, or just ***lead time***. And no, this isn't necessarily the sum of all the cycle times. This isn't a special run on Saturday for the CEO's birthday party, *when no other brownies are in the system*. Lead time tells you how long it takes one set of ingredients to go all the way through *when the factory is running normally*—and there's a difference. More on this later.

2. The question could also be "How often does a full batch of brownies come out of the system?" In other words, what's the cycle time for the whole process?

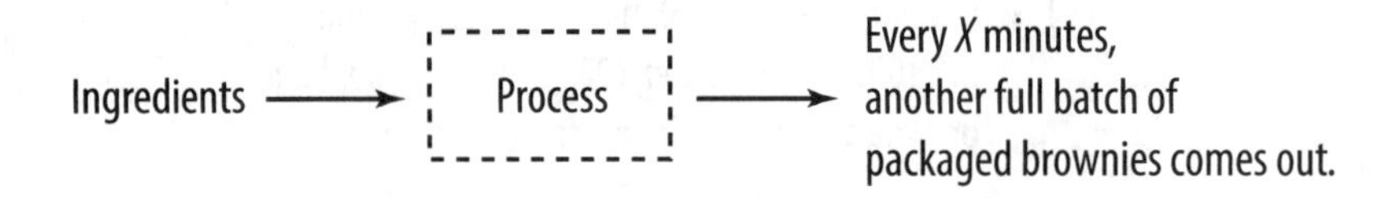


Look at the second question first:

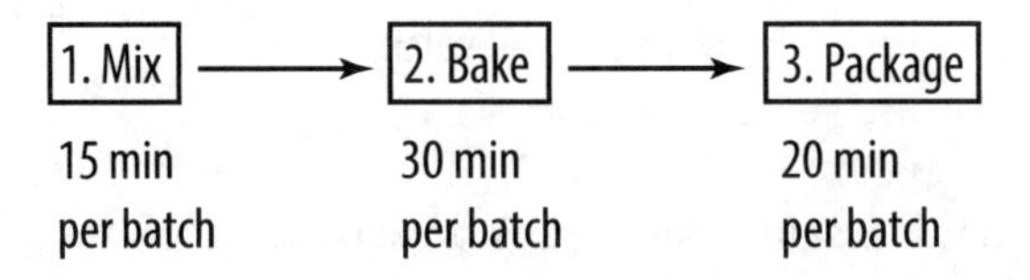


If the stages have different cycle times, then the overall pace is set by the ***slowest*** stage, which in this case is the Bake stage. If the mixers work constantly, they'll just pile up unbaked batter next to the ovens. And the packagers can't work continuously either—a new batch of brownies comes out of the oven every half hour, so the packagers do their thing for 20 minutes, finish the job, then sit around for 10 minutes of ***idle time*** as they wait for more brownies to be baked.

Here's the work schedule:

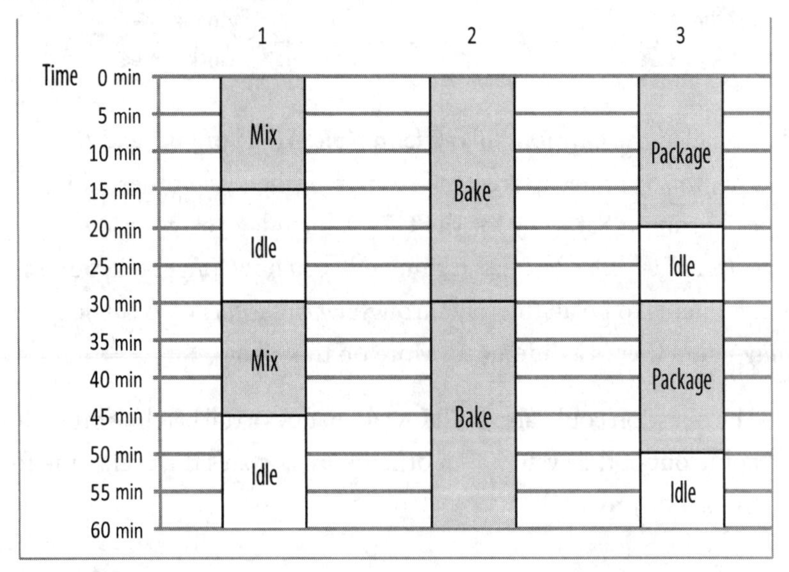

So the ***overall cycle time*** of the process is 30 minutes per batch, meaning that every 30 minutes, another full batch of packaged brownies emerges from the factory.

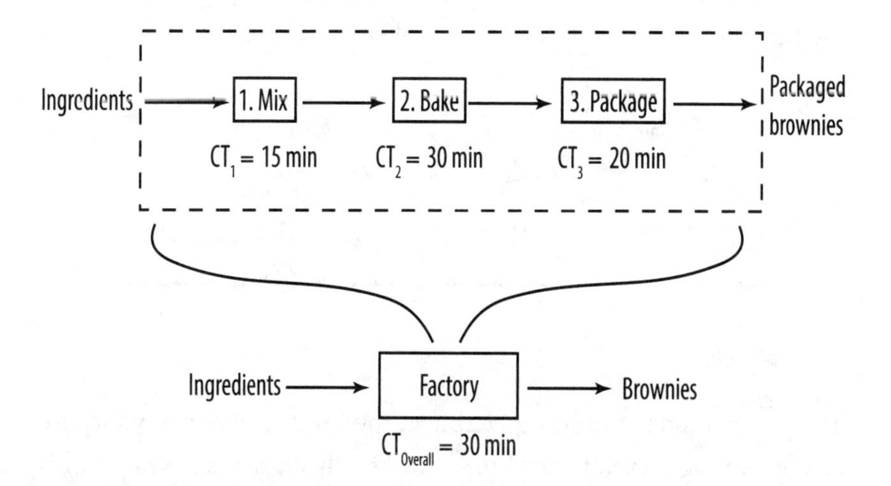


The Bake stage is what constrains your operation—it's your ***rate-limiting step*** or ***bottleneck***. If you owned the brownie factory, where should you spend more money?

- Hire more mixers and buy bigger mixing bowls?
- Buy a second oven, doubling your baking capacity?
- Hire more packagers?

Of course, you'd buy another oven. Spending more money on the other steps only increases the idle time. On the other hand, with another oven, you can't bake a single brownie any faster, but you can double the batch size at that step—so your effective cycle time for the Bake stage would be cut in half, from 30 minutes to 15 minutes. In that case, the Bake stage would no longer be the bottleneck. Instead, the slowest step would be Packaging (CT = 20 min).

The idea that the slowest stage sets the pace for the whole process was the point of an odd novel about operations that we were both assigned in school. This novel is called *The Goal,* and years later we both easily recall the somewhat offensive image of Herbie, the overweight kid on the Boy Scout hike who slows the whole group down:

The group winds up putting Herbie at the front, where everyone else can "encourage" (i.e., berate) him. Eventually, they relieve him of his backpack, lightening his load so they can all walk faster. All that's left behind is Herbie's dignity:

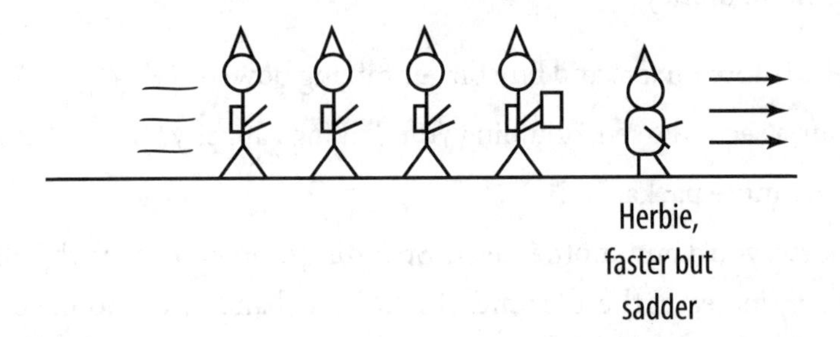


Neither of us likes this example, but at least it's memorable. If you want to speed up a process, find the Herbie—the bottleneck—and put all your efforts into making that part of the process go faster.

Back to brownies. Assume you haven't bought a second oven, so the process still looks like this:

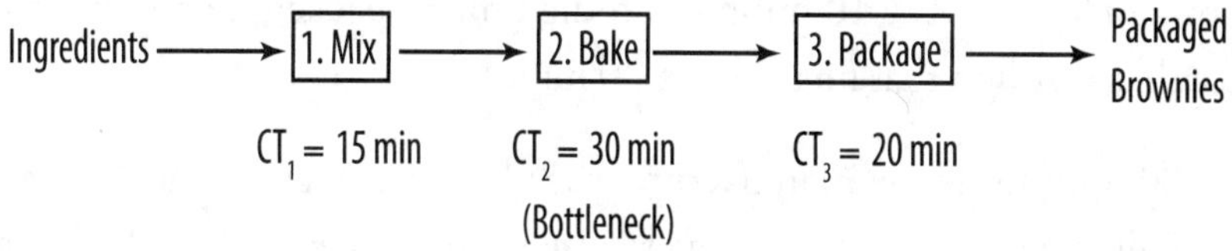

Ingredients → 1. Mix → 2. Bake → 3. Package → Packaged Brownies

$CT_1 = 15$ min  $CT_2 = 30$ min (Bottleneck)  $CT_3 = 20$ min

The overall cycle time is 30 minutes per batch. You can also say that the overall ***throughput time (TPT)*** is one batch per 30 minutes.

$$\frac{\text{Cycle time}}{\text{CT}} = \frac{30 \text{ min}}{1 \text{ batch}} \qquad \frac{\text{Throughput}}{\text{TPT}} = \frac{1 \text{ batch}}{30 \text{ min}}$$

Throughput is a rate of *stuff per time* (stuff/time), say batches/min. In contrast, cycle time is *time per stuff* (e.g., 30 min/batch). Cycle time and throughput are just reciprocals of each other.

$$\text{CT} = \frac{1}{\text{TPT}} \qquad \text{TPT} = \frac{1}{\text{CT}}$$

How long will it take a *particular* set of ingredients to get through the system under normal conditions? This is a different question: the question of lead time. Remember that all three stages, including idle time for the first stage, have to take 30 minutes per batch. However, you can chop off the last 10 minutes.

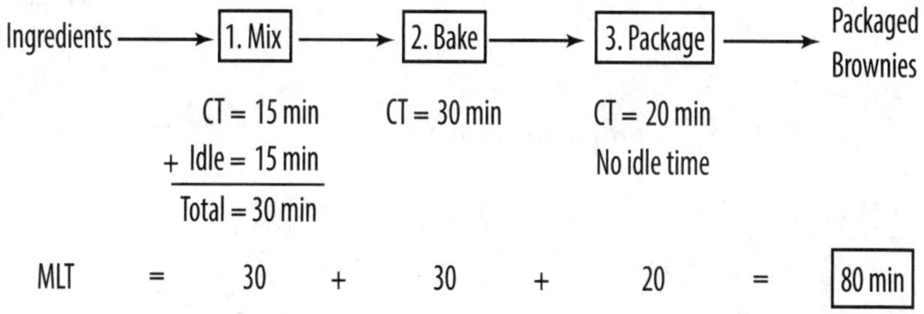

Ingredients → 1. Mix → 2. Bake → 3. Package → Packaged Brownies

1. Mix: CT = 15 min + Idle = 15 min; Total = 30 min
2. Bake: CT = 30 min
3. Package: CT = 20 min; No idle time

$$\text{MLT} = 30 + 30 + 20 = \boxed{80 \text{ min}}$$

You don't count the 10 minutes of idle time in packaging because the batch is through that final step in 20 minutes.

So 80 minutes is the manufacturer's lead time. Imagine that there are partially finished brownies at every stage in the assembly line. Now you decide to make a special batch of brownies with nuts.[5] If you add the nuts to the very next batch, how long will it take for that batch to get made? 80 minutes is the answer. Notice that you have to add in the first step's idle time to get the correct overall lead time because the special batch with nuts has to wait for the ovens to free up.

Be sure to keep the cycle time CT and the manufacturer's lead time MLT straight:

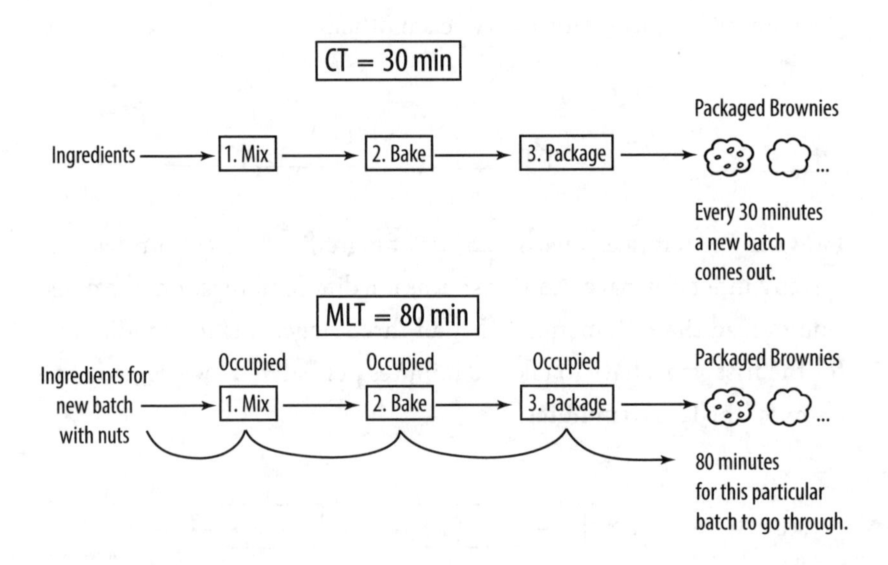


CT and MLT have a simple relationship:

$$\text{MLT} = 2\tfrac{2}{3} \times \text{CT}$$
$$80 = 2\tfrac{2}{3} \times 30$$

5 Or whatever your preferred additive is for brownies.

This 2⅔ is almost the same thing as three, the number of stages. One way to think of MLT is to multiply CT by the number of stages (three), then shave off the idle time from the final stage (10 minutes). That's why the number is slightly below three.

In fact, 2⅔ is the ***average number of batches in the system*** at any one time. The ⅔ comes from the fact that the last stage only takes 20 minutes (out of 30 for a cycle), so there's only a batch there ⅔ of the time. The batch is "gone" during the 10 minutes of idle packaging time, so we count just ⅔ of a batch in our average.

The 2⅔ batches being mixed, baked, and packaged are also known as ***work in process***, or ***WIP*** (pronounced "whip"). It's a form of unfinished inventory. The general formula is this:

| Manufacturer's Lead Time | = | Work in Process | × | Cycle Time |
|---|---|---|---|---|
| 80 min | = | 2⅔ batches | × | 30 min per batch |
| For 1 batch to travel through from start to finish | | The average amount of stuff being worked on at all stages | | The time between two batches coming off the line |

If a new batch comes out every 30 minutes, and it takes 80 minutes to go all the way through, there must be 80/30 = 2⅔ batches in the system. This relationship is known as ***Little's Law***. It works even when there's fluctuation in the system and the numbers are just averages. You can rewrite Little's Law in terms of throughput TPT, which equals 1/CT:

$$\text{MLT} = \text{WIP} \times \frac{1}{TPT}$$

$$\text{TPT} \times \text{MLT} = \text{WIP}$$

$$\left(\frac{1 \text{ batch}}{30 \cancel{\text{min}}}\right) \cdot (80 \cancel{\text{min}}) = 2\tfrac{2}{3} \text{ batches}$$

## QUEUEING

***Queueing*** theory is all about *waiting in line.* The next time you're at the DMV, just think: you can ponder this theory over and over and over again as you wait for your number to be called.

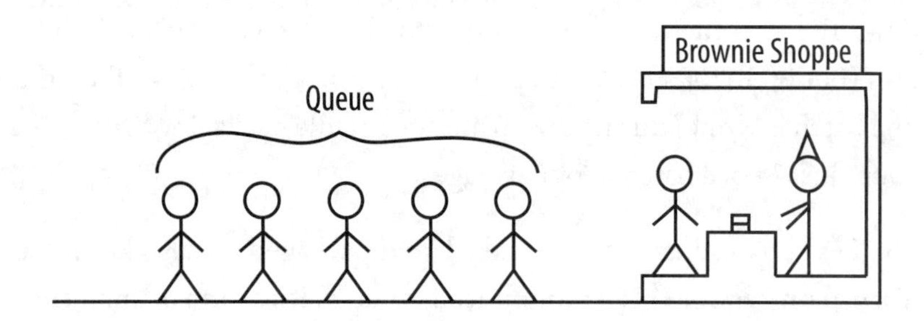


In the picture, there are five people in the ***queue*** (the line, to us Americans) and one person being served by the ***server*** behind the computer.

Little's Law shows up here too, if you think of the queue itself as a process. The people in the queue are like a work in process: unfinished inventory.

Say there are five customers on average in the queue—in other words, the ***mean queue length*** $L_q = 5$. Also say the ***mean wait*** $W_q$ in the queue is 20 minutes (that's just like lead time, the time for one batch to get all the way through the factory). What's the equivalent of cycle time? In other words, if someone is served every X minutes, what is X? Or rather, how often do you get to yell, "You got served!"?

*Process version of Little's Law:*

$$\text{Manufacturer's Lead Time} = \text{Work in Process} \times \text{Cycle Time}$$
$$\text{MLT} = \text{WIP} \times \text{CT}$$

*Queucing version of Little's Law:*

$$\text{Mean Wait in Queue} = \text{Mean Length of Queue} \times \text{Mean Service Interval between Customers}$$
$$W_Q = L_Q \times X$$
$$20 \text{ min} = 5 \text{ customers} \times X$$

$$X = \frac{20 \text{ min}}{5 \text{ customers}} = 4 \text{ min per customer}$$

This is often rewritten using a ***mean throughput rate R***, which would be one customer every four minutes in this case.

$$\text{Mean Length of Queue} \quad \text{Mean Wait in Queue} \times \text{Mean Throughput Rate}$$
$$L_Q = W_q \times R$$
$$5 \text{ customers} = 20 \text{ min} \times \frac{1 \text{ customer}}{4 \text{ min}}$$

This says that if you can serve one customer every four minutes and there are five customers in the queue on average, then each customer will wait a total of $4 \times 5 = 20$ minutes, on average.

This also works if you include the person being served:

| Mean Total People in System | = | Mean Wait in System | × | Mean Throughput Rate |
|---|---|---|---|---|
| $L$ | = | $W$ | × | $R$ |
| 6 customers | = | 24 min | × | $\frac{1 \text{ customer}}{4 \text{ min}}$ |

There are a few other symbols to know:

- ***Mean arrival rate* λ** (Greek lambda) = number of customers arriving every hour (or every minute, every day, or whatever time period you choose)
- ***Mean service rate* μ** (Greek mu) = number of customers being served every hour (or other time period)

Think of $\lambda$ as the mean rate of flow *into* the queue, while $\mu$ is the mean rate of flow *out of* the system.

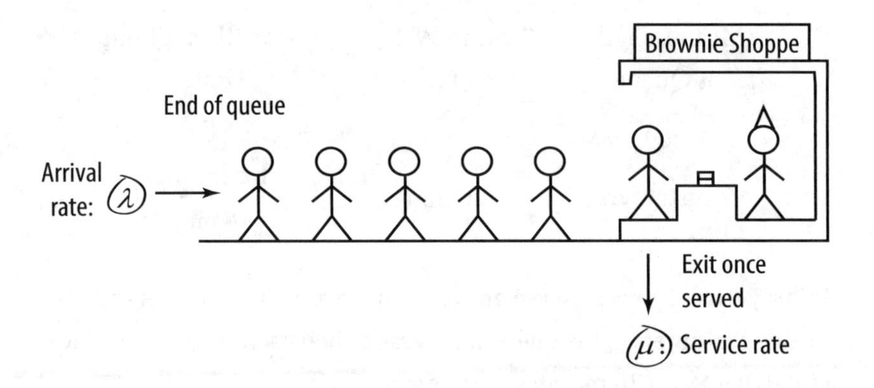


If the arrival rate and the service rate are the same and perfectly constant over time, then the line will be perfectly constant in length.

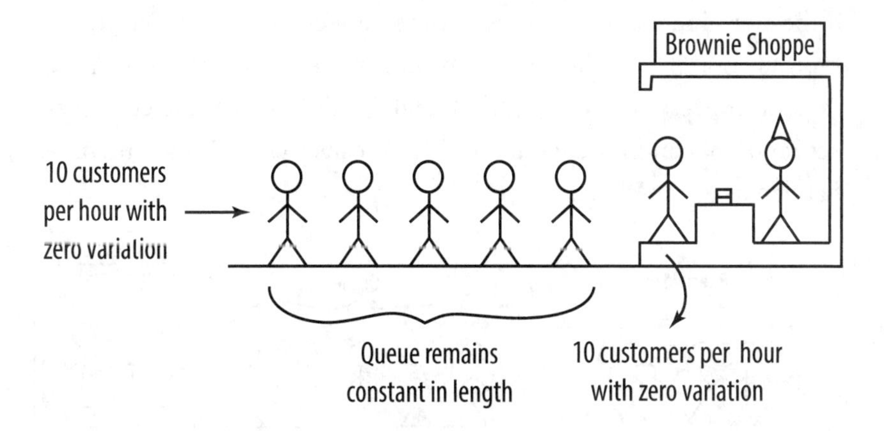


Notice, however, that *any* queue length works equally well here. You could have a constant queue length of 2 customers—or 20, or 200, and if the arrival rate and service rate are perfectly constant and perfectly matched, those queues won't change in length. But obviously, from the point of view of the shop, it's better to have a short queue than a long queue—you have happier customers when they wait less. No one wants to smell fresh-baked brownies they can't eat.

So how do you shorten the queue length? You need to have the service rate exceed the arrival rate.

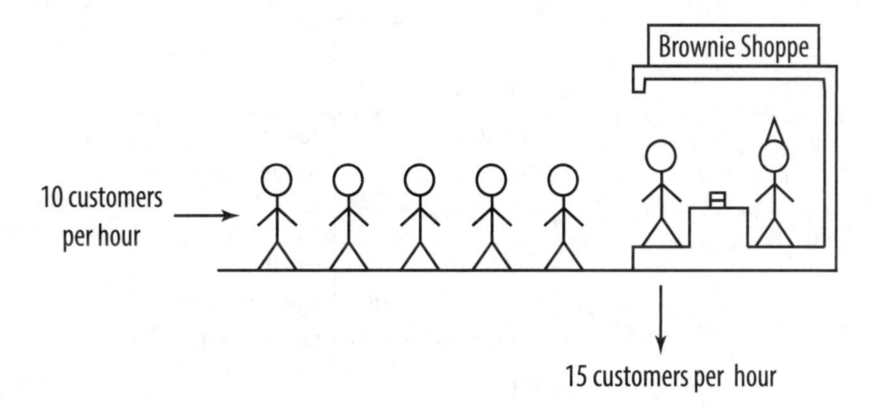


Logically, you'd think that this would drive the queue length to zero—it's like draining the tank faster than you're refilling it. You'd be right if the arrival rate and service rate were perfectly constant. But they're typically not. In fact, the arrivals and departures are usually modeled as Poisson processes (see Chapter 3: Statistics), in which the customers act randomly and independently. This means that arrivals sometimes happen close together...

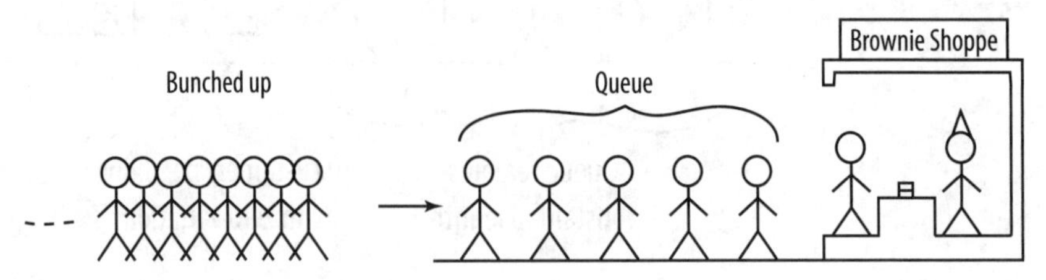


...and sometimes they're spread far apart.

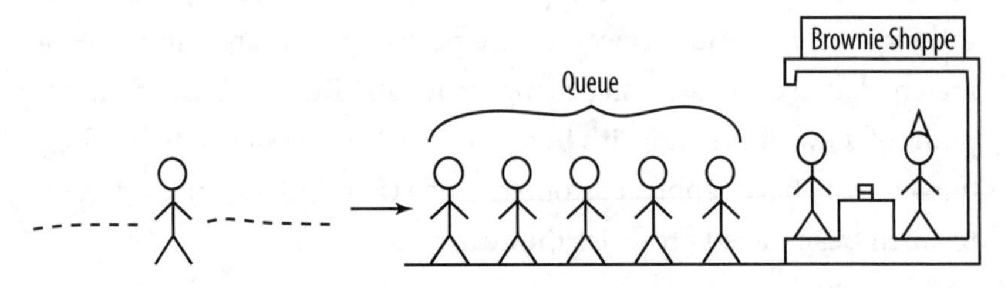


Departures (services) are similarly bunched—sometimes the order is quick, and sometimes it's slow.[6] Consider all the possible combinations:

| | | Service: Fast | Service: Slow |
|---|---|---|---|
| Arrival | Fast | Balance: queue remains same length | Faster arrival: queue grows |
| | Slow | Faster service: queue shrinks ***but cannot go below zero*** | Balance: queue remains same length |

6 Ever been to a Starbucks? Of course you have. What ever happened to ordering plain coffee? The guy in front of you has to have his grande soy extra-dry mocha latte with Splenda. That takes a fatiguing amount of time—both to order and to make.

You already know what happens when arrival and service are in balance—nothing. The queue stays the same length. When customers arrive faster than they're served, the line grows, and when customers are served faster than they arrive, the line shrinks. All that should make sense. Here's the key asymmetry, though—the queue can never get shorter than zero customers long. There's no such thing as a negative queue. What this means is that some of the *slow arrival + fast service* situations result in zero-length queues, and the server is idle. Unfortunately, you can't bank that idle time for the rush hours, when you have *fast arrival + slow service,* so the line tends to grow. Even if the arrival rate $\lambda$ equals the service rate $\mu$, any queue length from zero to infinity is equally likely if the arrivals and departures are randomly spaced. Over time, the queue length will explode.

This is very counterintuitive. It turns out that if you want a short queue, on average, then you'll want to have your service rate $\mu$ be *significantly faster* than the arrival rate $\lambda$.

$$\text{Good situation:} \quad \lambda = \frac{10 \text{ customers}}{\text{hour}} \qquad \mu = \frac{15 \text{ customers}}{\text{hour}}$$

The so-called ***traffic intensity*** (symbolized by the Greek letter $\rho$, or rho) is the ratio $\frac{\lambda}{\mu} = \frac{10}{15} = \frac{2}{3}$, in this case. You can also express this intensity in terms of the mean time between arrivals and the mean service time.

| Arrivals | Service |
|---|---|
| $\lambda = \frac{10 \text{ customers}}{\text{hr}}$ | $\mu = \frac{15 \text{ customers}}{\text{hr}}$ |
| $\frac{1}{\lambda} = \frac{1}{10}$ hr per customer | $\frac{1}{\mu} = \frac{1}{15}$ hr per customer |
| Mean time between customers = 6 min ($\frac{1}{10}$ hour) | Mean service time = 4 min ($\frac{1}{15}$ hour) |

$$\rho = \frac{\text{Mean service time}}{\text{Mean time between customers}}$$

$$\text{Traffic intensity } \rho = \frac{4 \text{ min}}{6 \text{ min}} = \frac{2}{3}$$

Unfortunately, the math on this shows a couple of troubling results.

1. The average ***utilization*** of your single server is equal to $\rho$.

   Joe behind the counter is only busy serving customers ⅔ of the time ($\rho = 2/3$).

2. The average queue length $Lq$ is equal to $\frac{\rho^2}{1-\rho}$.

   In this case, $L_q$ works out to $\frac{\left(2/3\right)^2}{1-2/3} = \frac{4/9}{1/3} = \frac{4}{3}$

   The line will be, on average, 1.33 customers long.

What if you want to reduce the fraction of time your server is idle from ⅓ to ⅕, so that $\rho = 4/5$? Then the average queue length becomes:

$$Lq = \frac{\rho^2}{1-\rho} = \frac{\left(4/5\right)^2}{1-4/5} = \frac{16/25}{1/5} = \frac{16}{5} = 3.2 \text{ customers in line}$$

That's letting your server be idle 20% of the time. What if it still drives you nuts to see him or her standing around? Say you only can tolerate 10% idle time. Then you'd need $\rho = 9/10$, and you'd have:

$$Lq = \frac{\rho^2}{1-\rho} = \frac{\left(4/10\right)^2}{1-9/10} = \frac{81/100}{1/10} = 8.1 \text{ customers in line}$$

And that's ***average*** line length—you'd often have many more waiting.

Long story short—if you want short queues for your customers, you need to have substantial "idle" time for your servers. They can of course be doing other things for you during that idle time, but don't expect them to be talking to customers. Queue length and server idle time are in tension, and you must trade them off against each other.

A couple of last things on queues—if you have multiple servers, it's generally better to ***pool the queues*** into one long line.

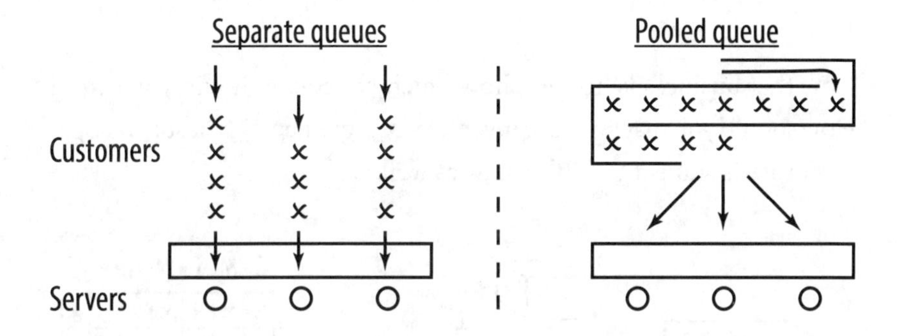


All kinds of businesses have caught on to this principle. Not only does it seem fairer to customers and reduce their stress of choosing the right line, a pooled queue often has shorter average waiting times, with less variance in those times as well.

However, the other principle is that a ***priority scheme*** for "quick" customers can help reduce average wait times for everyone. An express checkout line is good for you even if you aren't using it—it siphons off the quick-to-serve customers and reduces wait times for all.

## INVENTORY MANAGEMENT

Congratulations, you've optimized your brownie-making process and your customer queue.

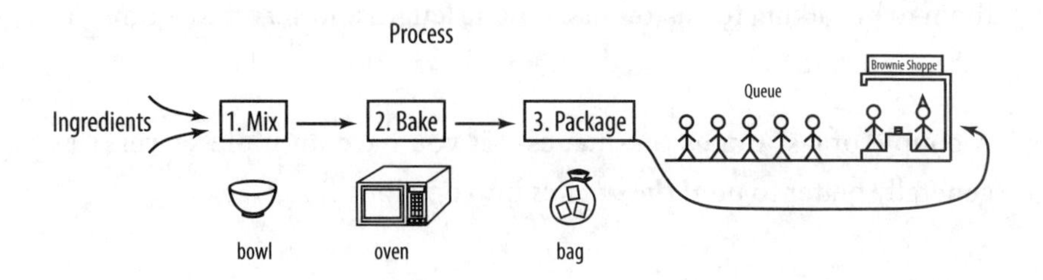


Now that business is hoppin', how much chocolate should you order? How should you manage your inventory of ingredients? Not surprisingly, inventory is subject to Little's Law as well.

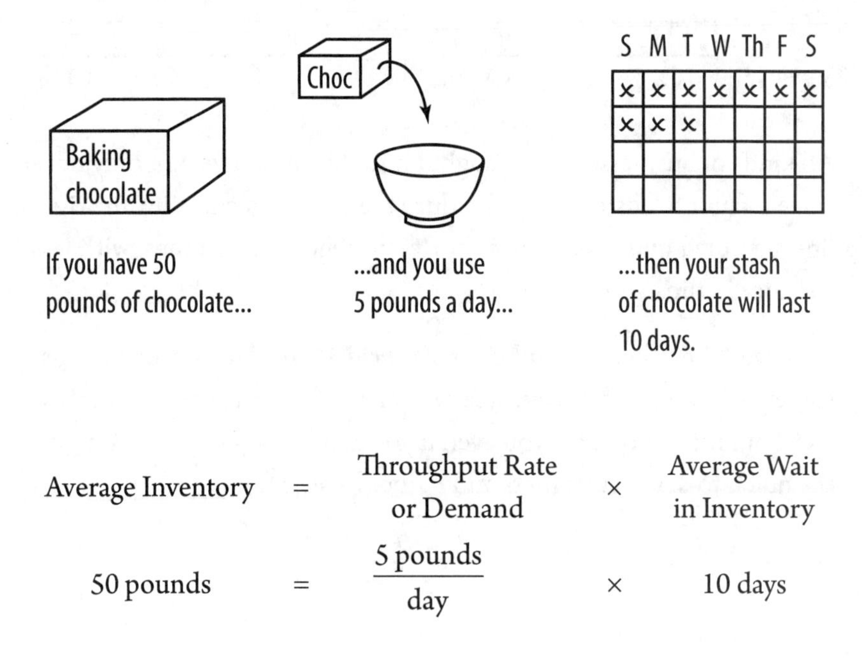


$$\text{Average Inventory} = \begin{matrix}\text{Throughput Rate} \\ \text{or Demand}\end{matrix} \times \begin{matrix}\text{Average Wait} \\ \text{in Inventory}\end{matrix}$$

$$50 \text{ pounds} = \frac{5 \text{ pounds}}{\text{day}} \times 10 \text{ days}$$

This section covers two fundamental models of inventory management. Both are simplified (like every other model in the world), but they provide important starting points.

1. Economic Order Quantity (EOQ)
   - Use when demand is perfectly constant over time.
2. Newsvendor
   - Use when demand is uncertain, and you only have one chance, in advance, to order inventory.

## ECONOMIC ORDER QUANTITY

Say you use 5 pounds of chocolate a day in your brownie shop—day in, day out. From now until the end of time you will use 5 pounds of chocolate every day. (This demand, or throughput, is labeled ***d***.)

There are two competing principles when it comes to figuring out your chocolate supply:

1. ***Order a big amount***

   This way, you save on the ***ordering cost***. This is *not* the cost of the chocolate itself. Every time you place an order, you incur a separate cost of, say, $45 between shipping, handling, receiving, and even placing the order itself. This ***per-order cost*** is usually labeled ***a***. You don't want to pay this cost too frequently, so you want to place fewer but larger orders.

2. ***Order a small amount***

   This way, you save on the ***storage cost*** (usually labeled ***h***, for holding). This cost includes rent on the storage facility, electricity for the refrigerator to keep the chocolate cold, and even spoilage of some fraction. All in, say this cost is $2 per pound per day, for every pound of chocolate you hold in inventory. This cost leads you to want to place frequent, small orders.

Conveniently, the minimum total cost comes when these two principles are in balance—that is, when the ordering cost and the storage cost equal each other over the course of time.

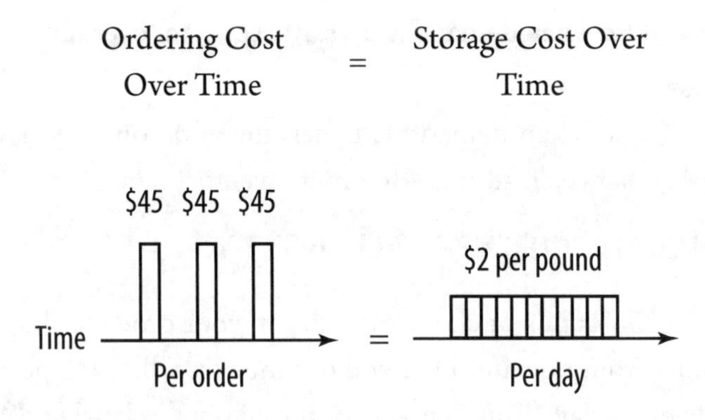


There is a "right size" for every order that balances these two costs. This right size is labeled *Q* (for quantity). To really grasp the principles at work, take two extreme examples:

***Example 1*: *Order Once Every 30 Days***

To meet the demand of 5 pounds of chocolate a day, you need to order $5 \times 30 = 150$ pounds each time. Over the course of 30 days, that drops steadily to 0, at a rate of 5 pounds a day:

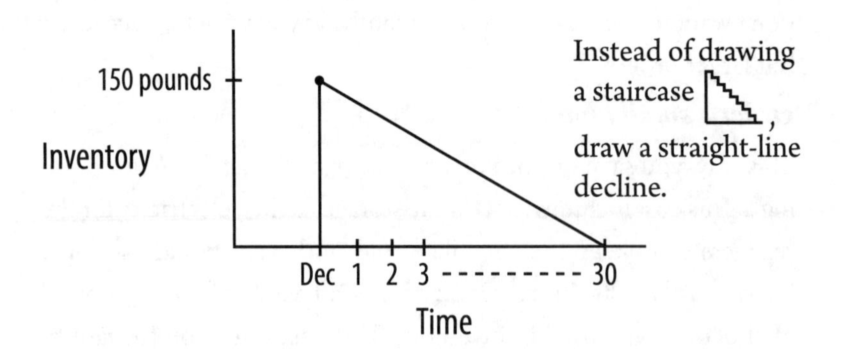


On December 30th, just as stocks reach 0, your next order of 150 pounds arrives. (Either your supplier can instantly meet your needs, like a genie, or you time your order perfectly in advance.) Another 30 days go by, the chocolate inventory falls steadily to zero, and the cycle repeats:

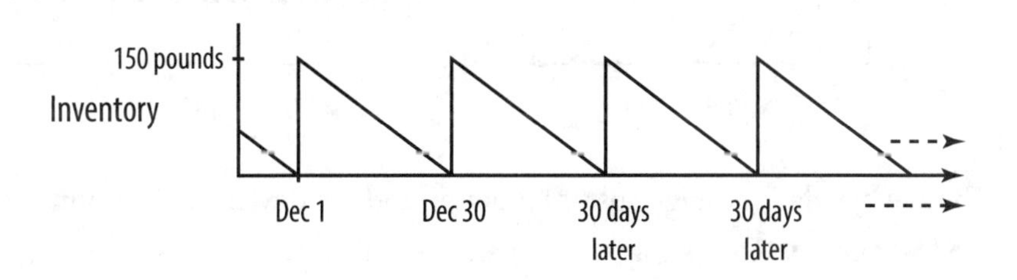


This sawtooth shape is typical of this sort of inventory model.

What are your costs over time (besides the cost of the chocolate itself, which will always be 5 pounds per day, times the price per pound)? In other words, what are your average daily costs of ordering and storing the chocolate?

1. ***Daily Ordering Cost = Low*** (because you're ordering infrequently)
   $45 per order every 30 days gives you a daily cost of
   $\$45 \div 30 = \$1.50$.

   You can write this using unit cancelation:

$$\frac{\$45}{\cancel{\text{order}}} \cdot \frac{1\ \cancel{\text{order}}}{30\text{ days}} = \frac{45}{30}\ \text{dollars}/\text{day} = \$1.50\text{ per day}$$

2. ***Daily Storage Cost = High*** (because you're keeping a lot around)
   Your inventory changes every day, but smoothly. You go from 150 pounds to zero at a steady rate, so you can just take the ***average inventory*** as $\frac{150+0}{2} = 75$ pounds.

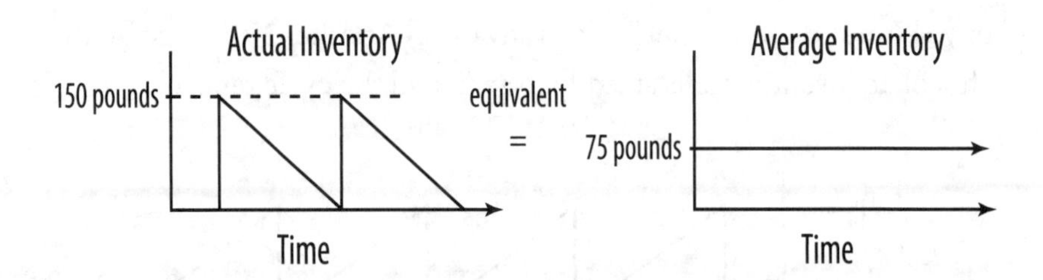


The average daily storage cost is \$2 per pound per day, times 75 pounds = \$150 per day. This is way more than the daily ordering cost of \$1.50, and the total daily cost of \$1.50 + \$150 = \$151.50 per day is too big.

***Example 2*: *Order Once a Day***

The other extreme is to order 5 pounds each day (since you use exactly 5 pounds per day).

1. ***Daily Ordering Cost* = *High*** (because you're ordering often)

$$\frac{\$45}{\text{order}} \cdot \frac{1 \text{ order}}{1 \text{ day}} = \$45 \text{ per day}$$

2. ***Daily Storage Cost* = *Low*** (because you're keeping very little around)

   The average inventory is ½ of the order amount, so the average inventory is only ½ × 5 pounds = 2.5 pounds.

   \$2 per pound per day × 2.5 pounds = \$5 per day

The total daily cost is lower than before (\$45 + \$5 = \$50), but you can go still lower when the two costs equal each other.

Take a look at the symbols again:

| | | | | |
|---|---|---|---|---|
| $d$ | = | demand, in pounds per day | = | 5 pounds per day |
| $a$ | = | ordering cost, per order | = | $45 per order |
| $h$ | = | storage cost, per pound per day | = | $2 to store 1 pound for 1 day |
| $Q$ | = | order quantity in pounds | = | how much to order each time |

Use $t$ for the days between orders. In the first case, $t$ was 30 days, and $Q$ was $5 \times 30 = 150$ pounds. The general formula is a rate-time-work relationship:

$$Q = d \times t$$

$$\text{amount} = \text{demand} \times \text{time}$$

$$150 \text{ pounds} = \frac{5 \text{ pounds}}{\text{day}} \times 30 \text{ days}$$

The daily ordering cost is the per-order cost ($a$, or $45) divided by the days between orders.

$$\text{Daily Ordering Cost} = \frac{a}{t}$$

Since $Q = d \cdot t$, you can solve for $t$ and substitute:

$$t = \frac{Q}{d} \rightarrow \text{Daily Ordering Cost} = \frac{a}{Q/d} = \frac{ad}{Q}$$

The daily storage cost is the "per-pound-per-day" cost, which is $h$, times the average inventory, which is $Q/2$.

$$\text{Daily Storage Cost} = h \times \frac{Q}{2} = \frac{hQ}{2}$$

Now set the daily ordering cost and the daily storage cost equal to each other, and lo—you'll get Q in terms of the other numbers.

$$\text{Daily Ordering Cost} = \text{Daily Storage Cost}$$

$$\frac{ad}{Q} = \frac{hQ}{2}$$

$$2ad = hQ^2$$

$$\frac{2ad}{h} = Q^2$$

$$\boxed{\sqrt{\frac{2ad}{h}} = Q}$$

In this case, we get $\sqrt{\frac{2(\$45)(5)}{(\$2)}} = 15$ pounds.

This is the ***economic order quantity***, or ***EOQ***. You should order 15 pounds at a time, which means you'll need to order every three days.

Test this quantity and see that you get the smallest total cost, compared to the total cost for any other order quantity.

$$\text{Daily Ordering Cost} = \frac{ad}{Q} = \frac{\$45 \cdot 5 \text{ pounds/day}}{15 \text{ pounds}} = \$15/\text{day}$$

$$\text{Daily Storage Cost} = \frac{hQ}{2} = \frac{(\$2 \text{ per pound})(15 \text{ pounds})}{2} = \$15/\text{day}$$

$$\text{Total Daily Cost} = \$15 + \$15 = \$30$$

With calculus, you can prove that this is the smallest possible cost—or you can just take our word for it. The principle that the costs should be the *same* to you both ways is rather deep, though—it's another aspect of the indifference principle. The best result often comes when you are indifferent between two options, or in this case, two costs.

At the end of the day, if you have constant, predictable demand on your inventory, then use this model. Figure out your per-order cost and storage cost, plug into the EOQ formula, and that's how much to order every time.

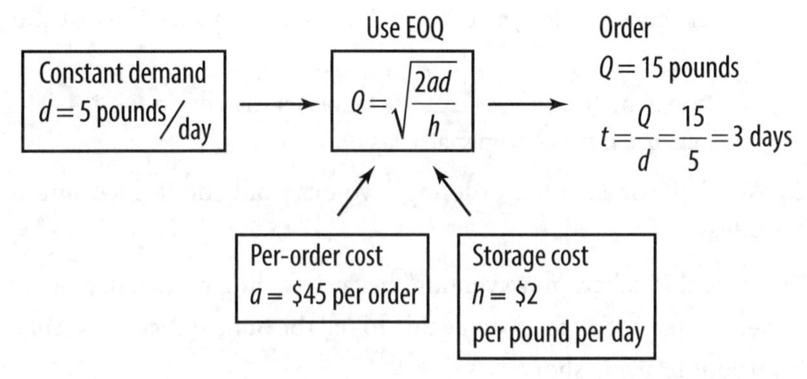


## NEWSVENDOR MODEL

The other fundamental inventory model involves very different assumptions.

1. Demand is now *random*. You don't know what it's going to be. All you know is that it will follow some *probability distribution*.[7]

7 See Chapter 3: Statistics.

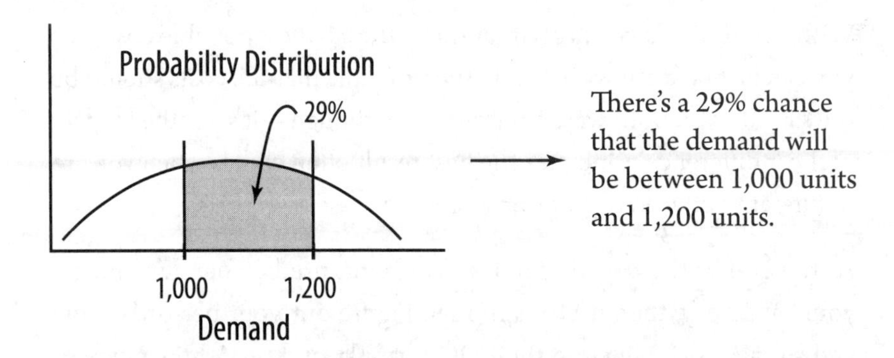


- That 29% was just completely made up as the area under the curve between 1,000 units and 1,200 units. The total area under the curve is always 100% (= 1). You usually take the distribution to be continuous rather than discrete because it makes the math easier.

2. You have one shot at ordering—you can only do it once, and in advance.

The model is called "newsvendor" because selling newspapers is the perfect scenario. Say that you decide to sell the Sunday *New York Times* from your brownie shop.

You have to order the newspapers the night before, but you don't know exactly how many you'll sell the next day—and you can't place a second order.

Some numbers can help make things more concrete:

- Cost per paper $c = \$2$
- Sale price of paper $s = \$5$
  (Ops folks use *s* rather than *p*, which could indicate probability.)
- Profit per paper $= s - c = \$3$

For now, assume that the papers are worthless if you don't sell them by the end of the day. In other words, the salvage value $v = \$0$.

There is a probability distribution for the demand, but you don't need to know the distribution's exact shape right now.

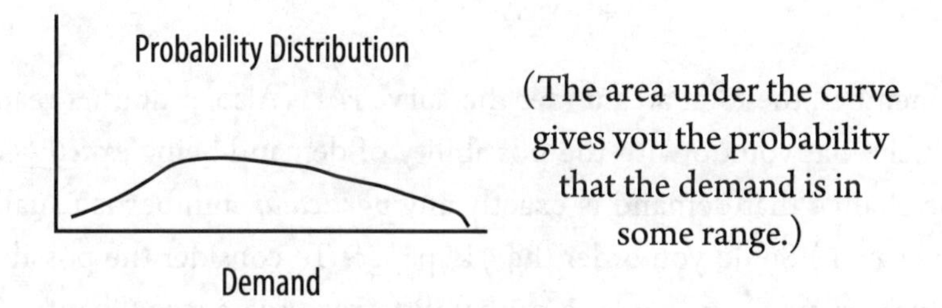


How many papers should you order on Saturday night? Imagine that you're already decided to order 40 papers, which happens to be way on the left side of the distribution:

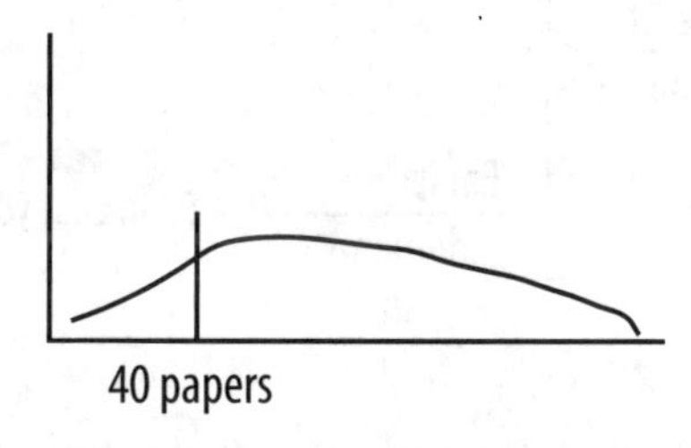


What's the chance that actual demand will be less than 40 papers? It depends on the shape of the distribution; say the answer is 15%.

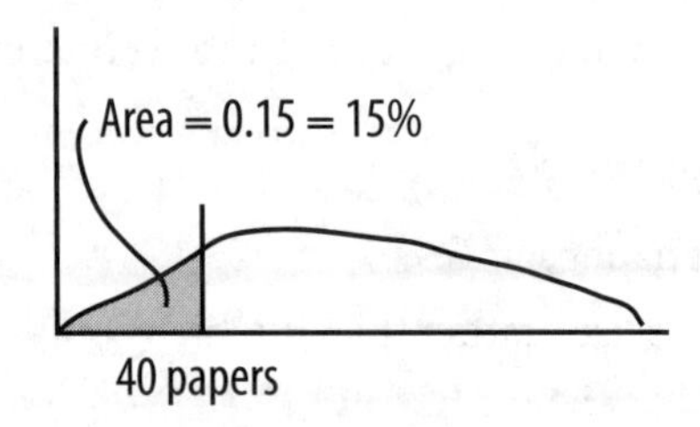


Then the chance that demand will be at least 40 papers is 85%.

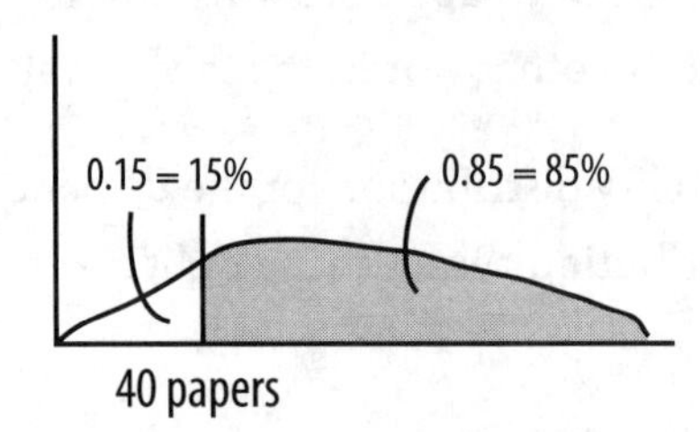


Remember, the total area under the curve is 1. (Also, it doesn't really matter what you do with the possibility of demand being *exactly* 40. The chance that demand is exactly any *particular* number is usually very small.) So do you order the 41st paper? To consider the possible outcomes, you can look at a probability tree. Again, see Chapter 3: Statistics for a refresher.

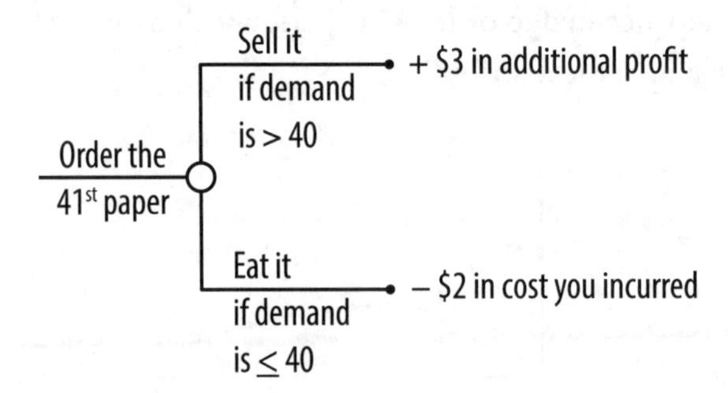


The $3 on the top branch equals $s - c$. It also has another name: ***underage cost*** (pronounced "UN-der-ij," not "under-AGE," which can result in different kinds of costs altogether). Underage cost, or $\mathbf{C}_u$, is the *cost of underestimating* demand by one paper. In this case, if you underestimate demand and you *could* have sold one more paper, you'd miss out on \$3 of profit, so $C_u$ = \$3. Likewise, ***overage cost*** ("OH-ver-ij"), or $\mathbf{C_o}$, is the *cost of overestimating* demand and "eating" one of your newspapers. Since the paper cost you \$2 and you get nothing for it (\$0 salvage value), $C_o$ = \$2.

What are the probabilities down each branch? Look at your distribution. Again, ignore what happens at exactly 40 papers.

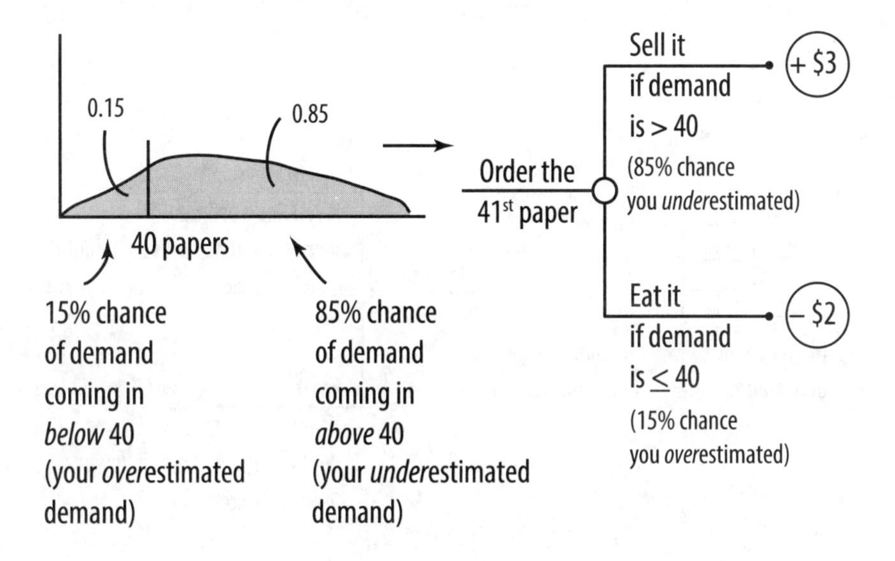


Would you take that bet? You have an 85% chance of winning \$3 or a 15% chance of losing \$2. So the expected value of the tree is:

$$
\begin{aligned}
\text{Expected Value} &= \left(\begin{array}{c}\text{Probability}\\ \text{of Top}\\ \text{Branch}\end{array}\right) \times \left(\begin{array}{c}\text{Value}\\ \text{of Top}\\ \text{Branch}\end{array}\right) + \left(\begin{array}{c}\text{Probability}\\ \text{of Bottom}\\ \text{Branch}\end{array}\right) \times \left(\begin{array}{c}\text{Value}\\ \text{of Bottom}\\ \text{Branch}\end{array}\right)\\
&= (0.85) \times (\$3) + (0.15) \times (-\$2)\\
&= \$2.55 + (-\$0.30)\\
&= \$2.25
\end{aligned}
$$

The expected value is positive, so you should take that bet and order the 41st paper, and likewise the 42nd, and the 43rd … Each one has a positive expected value to you, meaning that on average, you'd make money ordering that paper.

As you decide to order more and more papers, though, you move to the right on the distribution, and the probabilities of each branch shift. The chance of the top *good* branch falls, while the chance of the bottom *bad* branch rises.

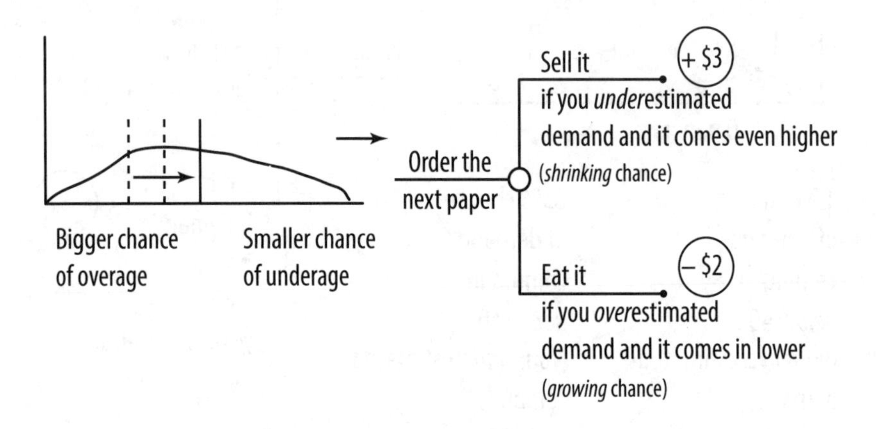


So the expected value of each additional paper is smaller and smaller, although still positive (so you decide to order it). At a certain point, though, the expected value of the next paper hits zero. At this magic number *Q*, the top branch's expected value (*EV*) cancels out that of the bottom branch.

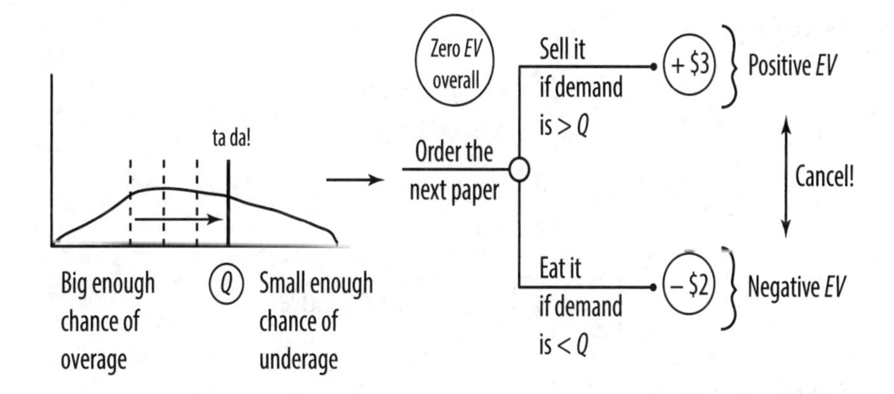


Beyond that point, additional papers *lose* you money—the trees have negative expected values. The chance of loss is too great.

So where is that magic *Q*? The answer lies in the probabilities:

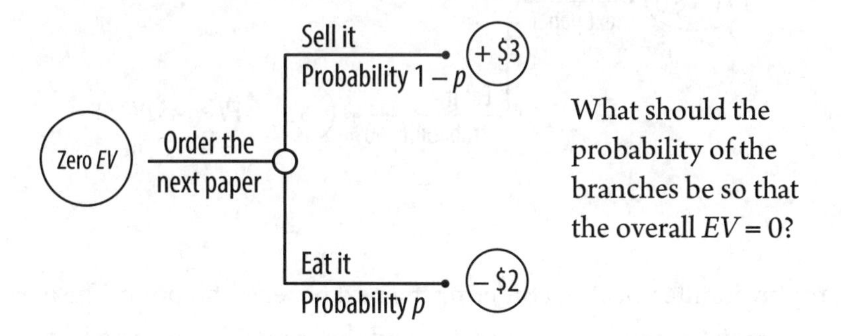


First, notice that if the bottom branch has probability $p$, then the top branch has probability $1 - p$. Now plug into the equation for the overall expected value:

$$\text{Expected Value} = \begin{pmatrix}\text{Probability}\\ \text{of Top}\\ \text{Branch}\end{pmatrix}\begin{pmatrix}\text{Value}\\ \text{of Top}\\ \text{Branch}\end{pmatrix} = \begin{pmatrix}\text{Probability}\\ \text{of Bottom}\\ \text{Branch}\end{pmatrix}\begin{pmatrix}\text{Value}\\ \text{of Bottom}\\ \text{Branch}\end{pmatrix}$$

$$0 = (1 - p) \quad (\$3) \quad + \quad (p) \quad (-\$2)$$

Solve for $p$.

$$0 = (1-p)3 - 2p$$
$$0 = 3 - 3p - 2p$$
$$0 = 3 - 5p$$
$$3 = 5p$$

$$\boxed{3/5 = p} \quad \text{so } p = 0.6 = 60\%$$

Put this on the tree:

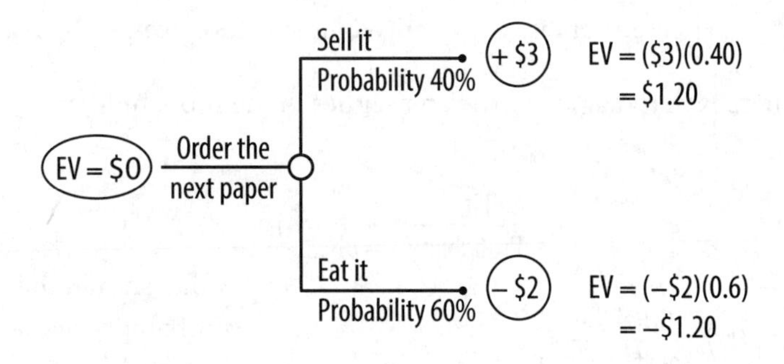


You are indifferent about ordering the next paper at this point. The *marginal profit*[8] is zero, and you've reached the maximum expected profit. Another way to look at the situation is on the distribution curve:

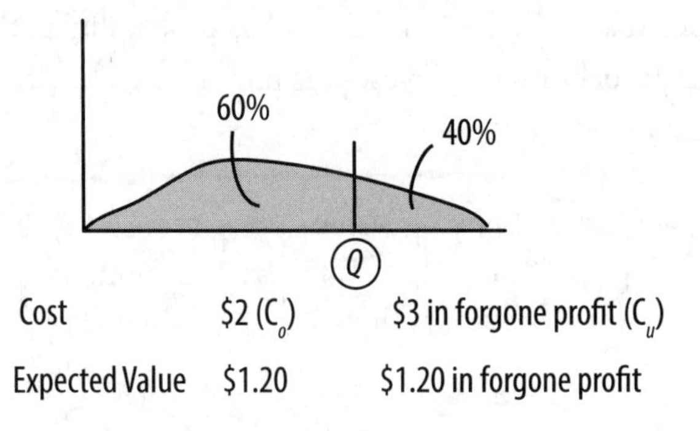


8 See Chapter 2: Economics & Game Theory for a refresher if needed.

The probability of overage $p$ (= 60%) is related to the overage and underage costs in the following way, putting in $C_u$ for \$3 and $C_o$ for \$2:

$$\text{Expected Value} = \begin{pmatrix}\text{Probability}\\\text{of Top}\\\text{Branch}\end{pmatrix}\begin{pmatrix}\text{Value}\\\text{of Top}\\\text{Branch}\end{pmatrix} = \begin{pmatrix}\text{Probability}\\\text{of Bottom}\\\text{Branch}\end{pmatrix}\begin{pmatrix}\text{Value}\\\text{of Bottom}\\\text{Branch}\end{pmatrix}$$

$$0 = (1-p) \quad C_u \quad + \quad (p) \quad -C_o$$

$$0 = C_u - pC_u - pC_o$$

$$0 = C_u - p(C_u + C_o)$$

$$p(C_u - C_o) = C_u$$

$$\boxed{p = \frac{C_u}{C_u + C_o}} = \frac{\$3}{\$3 + \$2} = 60\%$$

This ratio $\frac{C_u}{C_u + C_o}$ is called the ***critical ratio***. The magic order quantity $Q$ is uncovered by finding the point on the demand distribution at which the cumulative probability up to that point equals the critical ratio.

To recap, imagine that you have these costs in a newsvendor situation:

Underage cost (missed profit) = \$3 = $C_u$

Overage cost (unsold profit) = \$2 = $C_o$

What's the right order quantity?

a. *Draw the demand distribution.*

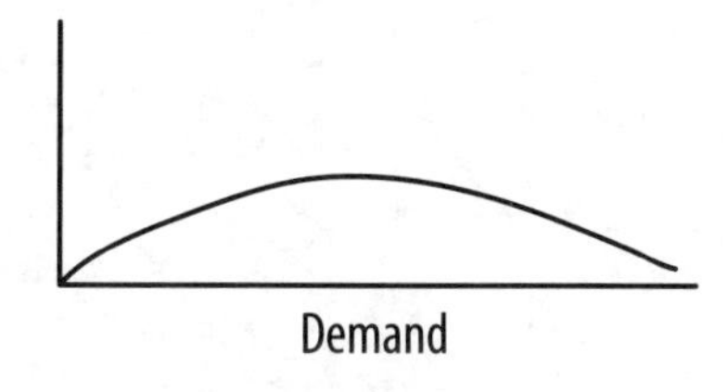

*b. Calculate the critical ratio.*

$$\frac{C_u}{C_u + C_o} = \frac{\$3}{\$3 + \$2} + \frac{\$3}{\$5} = 0.60 = 60\%$$

*c. Figure out the magic point Q where you've got 60% of the area on the left.*

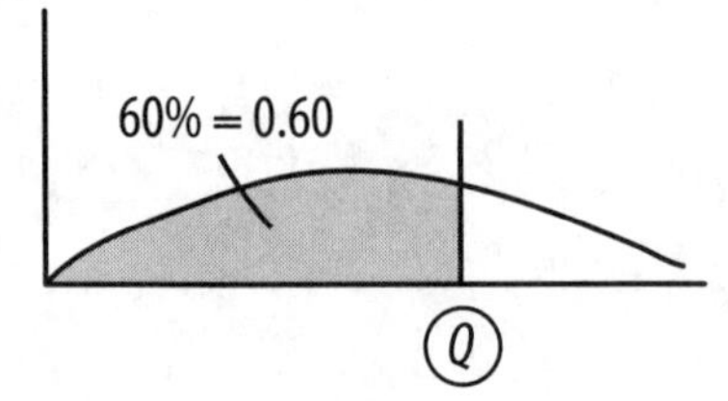


That's your answer!

If the $C_u$, the cost of being under the real demand, gets huge, then the critical ratio grows toward 100%. What if you could sell each paper for $20? Then you'd order lots and lots of papers.

$$\begin{aligned} s &= \text{sale price} &&= \$20 \\ c &= \text{cost} &&= \$2 &&= C_o \\ s - c &= \text{profit} &&= \$18 &&= C_u \end{aligned}$$

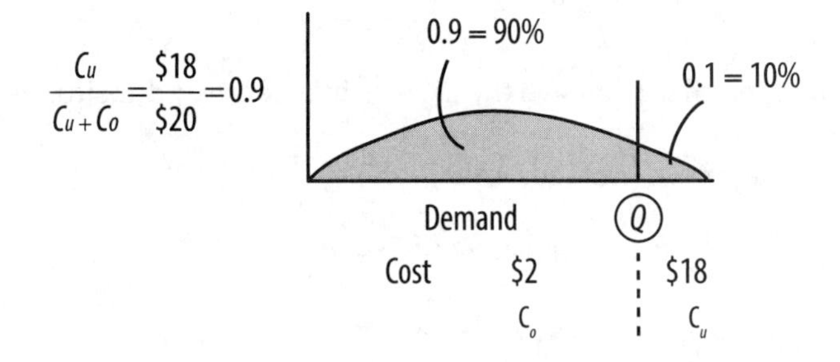

But if you could only sell each paper for $2.50, you'd order many, many fewer:

$$\begin{aligned} s &= \text{sale price} & & = \$2.50 \\ c &= \text{cost} & = \$2 \quad & = C_o \\ s - c &= \text{profit} & = \$0.50 \quad & = C_u \end{aligned}$$

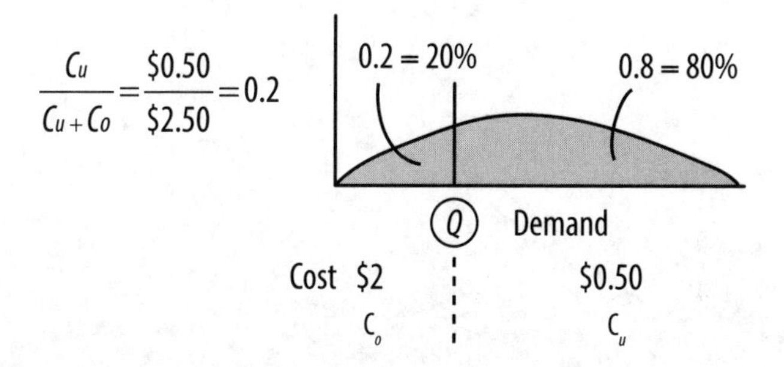


Lastly, if your goods have salvage value, you trim your overage cost. If you order too many papers, you won't end up losing as much money. So go ahead, order a bunch. Newspaper publishers will thank you.

# Chapter 8: Common Threads

As you go through the first year of school, people occasionally try to weave everything together for you—in a capstone strategy course, in an interdisciplinary business simulation, etc. However, you may still be left with the feeling that you've learned a lot of disconnected stuff.

This brief wrap-up will not completely eliminate that feeling. After all, you *do* learn a lot of disconnected stuff in business school. Still, it's worth highlighting a number of simple, repeated themes.

## SEVEN COMMON THREADS

The first three threads relate to the ***marketplace***. The fourth thread has to do with ***scale effects***, and the last three are about ***uncertain outcomes***.

> *Thread #1:* **Trade can benefit both sides of an exchange. The more different the two sides are, the more benefit they can extract from trade.**
>
> *Economics, Decision Analysis, Strategy, Negotiations*

Differences are what "grow the pie." If Ahmed really loves crust and you really love filling, then the two of you should figure this out and split the pie up so that he gets more crust and you get more filling. Granted, you aren't literally expanding the pie, but you're both happier than you would have been otherwise. This is why economists love trade: it can be so beneficial.

Differences in preferences (e.g., for the components of pie) and in endowments create opportunities for trade. But differences in *forecasts* or in risk tolerances can also create value. Every day, people happily take different sides of bets, whether in Vegas or on Wall Street. The reason is that people simply don't predict the same outcomes with the same likelihoods.

In fact, if a negotiation stalls, you might be able to break the logjam by adding in a bet. If you think your Widget 2012 is going to fly off the shelves, but Suzie Retailer won't stock it because she thinks it won't sell, then try to make the deal low-risk to Suzie on the front end. Consider paying her to stock the widgets; in return, ask for a bigger cut of any profits on the back end. This way, you're each putting your money where your mouth is. If the widgets don't sell, then Suzie makes out because she got cash up front. On the other hand, if the widgets sell like proverbial hotcakes, you get a bigger profit per unit than otherwise. In essence, you've taken on some of Suzie's risk for a greater possible upside.

As Mark Twain said, "It is difference of opinion that makes horse races."

*Thread #2:* **Competition in a marketplace forces down profits, because alternatives become easy to choose.**

*Economics, Finance, Marketing, Negotiations*

In the classic model of a perfect market, supply and demand interact to create a market price with zero "abnormal" profits. This model is introduced in microeconomics but shows up in practically every other subject. When you study a company's financial statements, for instance, you need to question abnormally high margins: why is this particular company able to earn such unusual profits? Was competition artificially restrained in some way? Are there large barriers to entry blocking potential new competitors? Is the product unique, so that the company functions somewhat like a monopoly?

Even when you aren't operating in a market, remember what competition does: it forces competition. In a one-on-one negotiation, improving your BATNA (best alternative to negotiated agreement) gives you more leverage. Your improved BATNA competes better with the other side's offer, and it becomes easier for you to walk away.

*Thread #3:* **Your competitor is as smart and as hardworking as you.**

*Economics, Finance, Strategy, Negotiations*

Even though you try not to underestimate the other side, you probably will—it's very hard to overcome the relevant cognitive biases. Small, active investors play the stock market thinking that they're smarter than average. But every time they buy or sell, someone who's probably more knowledgeable is on the other side of the transaction. In the long run, those active little traders tend to lose out, in some cases perfectly timing the market to buy high and sell low.

Consider taking systematic measures to give proper weight to the other side's point of view. For instance, set up a formal "devil's advocate" role on your team. A football team gets better when its practice squad effectively simulates the next opponent's strategy.

One of the best ways to learn to respect your competitor is to have your butt kicked a few times when the stakes are not so high. This is why simulations can be so useful. At the end of the day, there's no teacher like experience.

> *Thread #4:* **When you double a recipe, not everything about the recipe exactly doubles.**
>
> *Economics, Operations, Accounting, Marketing*

You don't bake two cakes for twice as long as you would bake one cake. Some elements of the recipe, such as baking time, increase by less than 100%—if they increase at all. When costs increase by less than 100%, you have *economies of scale.*

However, if you eat both cakes in one sitting, you don't enjoy the second one as much as the first one. In fancy terms, the marginal benefit of the second cake is less than that of the first cake. This situation is known as *diminishing returns,* decreasing returns to scale—or indigestion.

In both cases, the slope of the curve diminishes as you go out to the right:

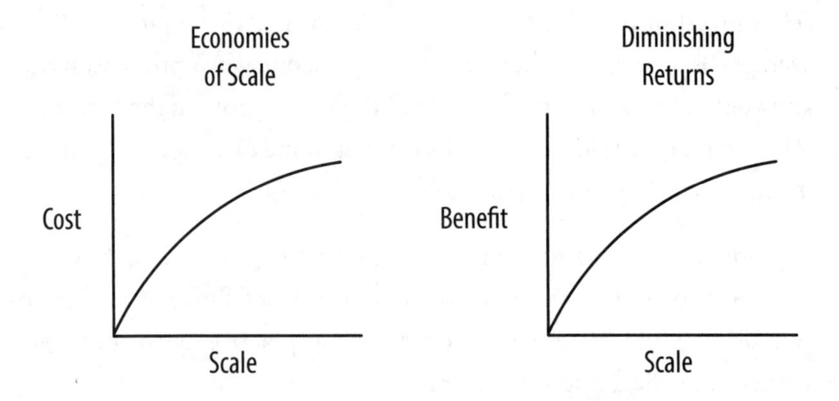


On the other hand, some parts of the recipe double exactly. For instance, you need twice as many eggs for two cakes as you do for one. Such components increase linearly with the scale of the project.

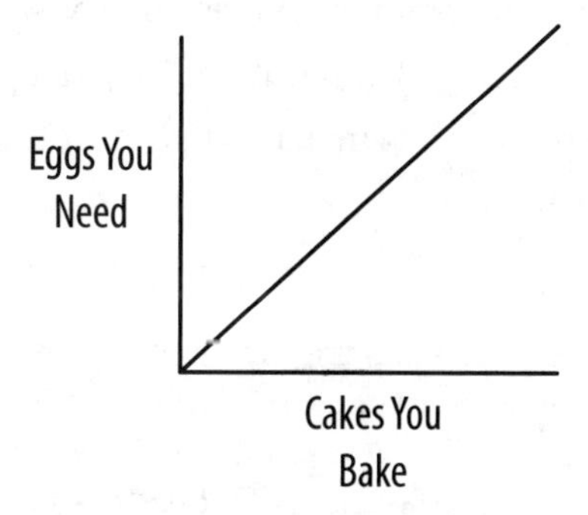


Finally, certain costs might suddenly jump as you increase the recipe. Maybe you want to start selling cakes as a business. You can't just multiply your recipe by 20 and buy 20 times as many eggs. You need a new kitchen, a delivery truck, several Keebler elves, etc. Costs that go up more than linearly are called *diseconomies of scale.*

On the other hand, you might encounter *increasing returns*. Perhaps now that you can guarantee a regular supply of cakes, you can sell them through Magnolia Bakery (which has all the bomb frostings), and you'll make much more per cake than ever before.

In both cases, the slope of the curve increases as you go out to the right.

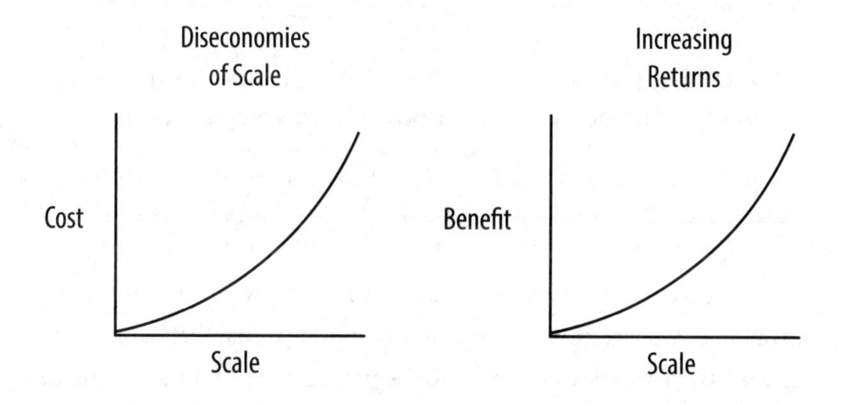

As you consider how an operation can be rescaled or what happens when more or fewer products are sold, think about what happens to important components—whether they change linearly or not, and if not, which way the curve bends and why.

> *Thread #5:* **You can assess the attractiveness of small bets with expected value.**
>
> *Economics, Statistics, Decision Analysis, Finance*

Here's a bet: you have a 70% chance of winning $5 and a 30% chance of losing $10. Should you take this bet? Yes, since the expected value is (0.70)($5) + (0.30)(–$10) = $3.50 – $3.00 = $0.50. In other words, if you take this bet over and over, you'll average a $0.50 profit per bet. Take this bet as often as you can (or let us do it).

Expected value is a great way to combine probabilities and outcomes into a single number. It's most appropriate as a criterion when the bet is small—in other words, when the worst case won't break you financially (assuming you're not allowed to declare bankruptcy and walk away).

> *Thread #6:* **You can assess the risk of a bet by considering the uncertainty and variability of the potential outcomes, as well as the correlation to other bets you've taken.**
>
> *Statistics, Decision Analysis, Finance, Strategy*

One aspect of risk is *uncertainty*: the fact that an outcome is not perfectly guaranteed. Uncertainty can be quantified with probability, which varies between 0% (impossible) and 100% (guaranteed). In finance, the discount rate is often used to capture uncertainty—an uncertain future cash flow is assigned a higher discount rate than a guaranteed future cash flow. (As an aside, the guaranteed future cash flow would not have a discount rate of zero, but rather some positive "risk-free rate." The discount rate captures not only the uncertainty but also the delayed gratification of a future cash flow.)

In addition to uncertainty, another aspect of risk is the *variability* of possible outcomes. Two bets can have the same expected value but have vastly different variability:

| Bet #1 | Bet #2 |
|---|---|
| 50% chance of winning $5 | 50% chance of winning $5 million |
| 50% chance of losing $4 | 50% chance of losing $4,999,999 |
| Expected value: $0.50 | Expected value: $0.50 |

The second bet has a much greater variability, and you'd be much less likely to take it than Bet #1 (again, assuming that you can't declare bankruptcy). You can quantify this variability with standard deviation. The first bet has a standard deviation of $4.50, whereas the second bet has a standard deviation of nearly $5 million. For more complicated bets, you might separately consider the worst-case scenario before placing your chips.

Finally, you can assess the riskiness of a bet by considering how the outcomes are *correlated* with the outcomes of other bets you have already taken. For instance, insurance is a bet you take to reduce your overall risk. Your insurance policy offsets other bets that you take in life—simply by owning a house or being alive, say. You purchase fire insurance so that it pays you when the outcome of your "owning a house" bet is bad (that is, fire destroys your house). Notice that these outcomes are negatively correlated: you get a good result from the insurance bet when the house bet goes south.

In other cases, you might decide to increase your risk by taking on a bet that is positively correlated with your other bets. This kind of *speculation* is similar to doubling down at the blackjack table (you add to your bet when your hand is strong and the dealer is showing a crummy card).

*Thread #7:* **You should demand a higher expected return for taking on higher risk.**

*Economics, Stats, Decision Analysis, Finance, Strategy*

If you are asked to take on risk, you should ask to be compensated for it. A higher risk demands a higher expected return. Likewise, if you reduce your risk, say by sharing the investment with a partner, you should expect to share any upside as well. Of course, you can take dumb risks that no one will compensate you for, such as sticking your face in a fan, but that's obvious.

The connection between risk and return is hammered home throughout the b-school curriculum. What's often not emphasized enough is how you actually measure return (generally, by expected value or expected utility) and risk (generally, by assessing uncertainty, variability, and correlation with other bets).

## TWO KEY TENSIONS

*Key Tension #1:* **How rationally can you expect people to behave?**

There are two camps on this issue, which is a never-ending source of debate.

*Camp A*: **People are basically rational when money's on the line.**

This is the position of neoclassical economics. In fact, *homo economicus* is the abstract person in standard economic models who pursues his or her own self-interest perfectly and maximizes his or her own utility at all times. According to the Camp A view, markets generally work efficiently if left to their own devices. Camp A includes not only neoclassical economics but also standard finance theory, decision analysis, and so on.

*Camp B*: **Even when supposedly acting in their own self-interest, people are not as rational as Camp A thinks.**

The fields of social psychology, cognitive science, and organizational behavior have all contributed to this position, which identifies systematic, persistent biases in the way people perceive, decide, and act. According to this view, markets do not work as efficiently as Camp A imagines. Camp B thinking has given rise to the newer fields of behavioral economics and behavioral finance. Some Camp B folks simply resist the reduction of human behavior to utility maximization.

*Key Tension #2:* **How much should people spread their bets?**

This is a narrower debate academically than the first, but it affects your life directly. Again, there are two schools of thought.

*School 1*: **Spread your bets. Don't put all your eggs in one basket.**

This is the idea of a *portfolio play*. Pick lots of offsetting bets, and you'll ride safely through the storms. According to this view, options (both in stocks and in life) are always worth something, so you keep your options open. You see this position a lot in traditional economics and finance.

*School 2*: **Concentrate your bets. "Put all your eggs in the one basket and—*watch that basket*."** (Mark Twain)

This is the idea of *strategic commitment*. You can sometimes change the game for the better by putting all of your chips on one number. In this view, options can actually have negative value, so under certain circumstances you preemptively close them off. You see this position in game theory, where it has been developed the most.

Schools 1 and 2 can be reconciled, perhaps, if the scenarios are spelled out. It's often best to spread your bets when you don't have control over the outcomes—you're entrusting your baskets to lots of other people. On the other hand, when you have control and you want to bind your future self to a course of action, it may be better to commit yourself to a single basket. Some decisions are all-or-nothing, at least in a sequential sense (e.g., the choice of an MBA program, a job, or a mate). With these decisions, you are forced to put all your eggs in one basket, so you may as well commit to that basket wholeheartedly.

## CAPSTONE COMMENTS

Fortunately, you don't have to pass a general exam before they hand you a diploma. Unfortunately, that means no one will demand that you really pull together everything you've learned. Broad strategy courses aim to paint the big picture, but they often fall short of synthesizing the full range of topics encountered in business school. Part of the reason is that Strategy has its own particular material to cover. Moreover, as a generally less quantitative field, Strategy glosses over mathematical insights from other parts of the canon.

We think it's useful to consider how all the subjects in the b-school curriculum interact, as well as what they tell you about yourself. For instance, relative to the key tensions above, where do you stand? Which threads are most apparent to you? And what are the implications for your career choices? As you encounter these key tensions in your courses, think about how they might relate to you personally, to your own values and decisions. The more you connect the big-picture debates to your own life, the more meaningful your educational experience will be in business school.

If you've made it this far, congratulations! We hope that this introduction to important b-school concepts has been helpful. Please drop Chris a line with any suggestions at cryan@manhattanprep.com. Good luck with business school!

# Appendixes

## A. GLOSSARY OF COMMON TERMS

**Accrual** – A cruel, cruel accounting method wherein revenue and expenses are recognized when they are incurred, regardless of whether any cold, hard cash was involved. Accrual accounting is different from cash accounting, which follows the greenbacks.

**Acme** – The name given to any company in a simulation when the author is too lazy to come up with something more interesting. To its credit, Acme produces rocket sleds and superhero outfits for Wile E. Coyote.

**Acquisition** – When one company buys another. Or any time you acquire something... like your study partner's notes.

**Amortize** – To account for something over a long period of time. If you amortize the cost of your MBA over your entire lifetime, it's still crazy expensive.

**Angel** – An investor who comes in at the beginning stages of an enterprise and provides seed funding. Kind of like the classmate who saves your butt right before a midterm.

**Arbitrage** – Making a penny by buying a financial asset at $0.99 and instantly selling something equivalent for $1.00. Arbitrageurs try to do this over and over to make lots of pennies. Arbitrage is definitely not a French body spray.

**Balance Sheet** – A snapshot of a company's finances, including assets, liabilities, and shareholders' equity.

**Bandwidth** – What you're able to keep up with. Also what your Internet provider is operating on.

**Barriers to Entry** – The industry equivalent of a chastity belt. Barriers to entry such as high start-up costs keep potential competitors from joining the fray.

**Benchmark** – The standard by which something is judged. Or to measure against those standards. Once upon a time, someone measured something by marks on a bench, and the rest is word history.

**Beta** – The Greek letter B (β), used to indicate (1) an early version of a product made available to customers, (2) the slope of a regression line in statistics, or (3) the risk of a financial asset, when it is added to a market portfolio. The beta of the market portfolio itself equals 1, while the beta of a perfectly risk-free asset equals 0. Also, beta is a very popular type of pet fish.

**Bleeding Edge** – Where the technological leading edge goes bad.

**Blue Chip** – A term for large, important companies, generally those that make up the Dow Jones Industrial Average. The most valuable poker chip (helps to keep a few up your sleeve).

**Bond** – A formal loan to a company or government, repaid with interest.

**Book Value** – A company's accounting assets minus its liabilities, often much different from its market value. Book value is not, technically, what you can sell your texts to first years for.

**Bottom Line** – The profit of a company, so called because it's listed on the bottom line of an income statement.

**Brand Equity** – The value of all of the warm-and-fuzzies that people feel toward a particular product or company.

**Bricks-and-Mortar** – Anything that isn't prefaced with an *e-*; a physical presence.

**Buggy Whip** – A general term used to describe obsolescent technology (e.g., buggy whips after the automobile was invented). The term *buggy whip* itself is becoming old-fashioned, like *eight-track tapes* or *CD-ROMs.*

**Business Cycle** – The up-and-down swing of a national economy over the course of years.

**Business Model** – The way a company makes money or the reasons why it should.

**Business Plan** – The document that (1) you write and that (2) no one reads and that (3) tries to explain how your startup will make money and the reasons why it should.

**C-Level** – Refers to top-level executives, such as CEO, CMO, CTO. *C* is for "Chief."

**Capital** – Money gathered for investment.

**Cartel** – A group of sellers that work together to force prices higher than the market would otherwise dictate. OPEC is the cartel of oil-producing countries.

**Cash Cow** – What we would all love to have grazing in our pastures. A tried and true product that generates large profits without requiring much in the way of resources. Don't expect cash cows to live too long. Eventually, something new comes along to slaughter them.

**Commercial Paper** – Short-term loans to companies. These loans are not secured by any specific asset as collateral.

**Commodity** – Undifferentiated stuff that is bought and sold, such as oil, soybeans, or copper. If a market becomes commoditized, there's going to be a race to the bottom in price, because price becomes the only differentiating factor.

**Common Stock** – Represents ownership in a company.

**Comps** – Also known as comparables. These are the numbers from the previous year that you use to assess current performance. *Comps* can also refer to the analysis of comparable companies in a pitch book.

**Deck** – A group of PowerPoint slides that guides a discussion or tells a story. Also known as a pack.

**Deflation** – A general fall in prices in an economy. Sounds good. Is bad.

**Depreciation** – The decline in value of a fixed asset.

**Derivatives** – Side bets in the casino of Wall Street, used either to lower risk (hedge) or take on more risk (speculate). For both entertainment and education, watch Richard Thaler, who won the Nobel Prize in economics, and Selena Gomez explain derivatives in the must-see movie *The Big Short*.

**Discount Rate** – An effective rate of interest, used to *discount* future cash flows to their present values.

**Diversification** – Holding different financial instruments as a way of spreading investment risk.

**Economies of Scale** – The corporate version of Costco; savings that can be achieved by doing something in bulk.

**Elevator Pitch** – The uber-quick spiel by which you sell yourself, your idea, or your company while the elevator descends from Floor 19 to the lobby.

**Equity** – Ownership. In b-school, this term almost never means "fairness," perhaps not surprisingly.

**Excel** – The dominant spreadsheet program, although Google Sheets is coming up quick. Your best friend and worst enemy.

**Expected Value** – The long-run average value you'd get if you took a bet many, many times.

**Federal Reserve** – Also known as "The Fed." The central bank of the United States (really a system of banks), which controls the US money supply and acts like a super-bank to typical banks and to the US government. The Fed is run by a board of governors and a chairman, whose face shows up a lot in the *WSJ*. Not the same as "the Feds," i.e. federal agents.

**Federal Funds Rate** – The interest rate banks charge each other when they lend each other money overnight to meet reserve requirements. The Fed sets the target "fed funds" rate.

**Fiscal Year** – A 365-day time frame for a company. Fiscal years don't necessarily correspond to any calendar, whether Gregorian, lunar, or Mayan.

**Fixed Cost** – Whether you make one widget or one million, your fixed costs stay the same. Hence the term "fixed." Nice how that works, no?

**FOMO (*Fear of Missing Out*)** – The irrational fear that nearly all business school students have at some point that they are not doing enough of whatever their classmates are doing a lot of. After all, YOLO, right? (*You Only Live Once.*)

**Forward** – An agreement to trade something for money at a future date.

**Future** – A type of standardized forward contract traded on an exchange.

**Going Concern** – The assumption that a company is going to continue operating indefinitely and is not in liquidation. Not always a good assumption.

**Goodwill** – A kind of intangible asset recorded on the books after an acquisition. Contrary to its use in everyday life, the term *goodwill* means nothing fuzzy or nice in accounting, partly because there is not much fuzzy or nice in accounting.

**Hedge** – To take an offsetting bet as a form of insurance, lowering your risk. In speech, to *hedge* is to be evasive, as in "beating around the bush."

**Inflation** – A rise in the general price level in an economy. A little bit of inflation is good; a lot of it is really bad. Many indulgences seem to work this way.

**Initial Public Offering (IPO)** – The sale of shares of a formerly privately owned company to the general public. When the founders, angels, and venture capitalist investors get rich.

**Investment** – What you tell yourself these two years are, an investment in your future.

**Just-in-Time (JIT)** – An approach to manufacturing that matches production with need in order to reduce inventory carrying costs. Justin Timberlake also matches production with need.

**Karaoke** – Rarely as good an idea as you think.

**Keynes** – Economist who gave his name to Keynesian economics, which argues for significant government intervention in the economy to try to smooth out business cycles.

**Leverage** – The use of debt to amplify a bet. It's just like borrowing money to gamble at a casino: you'd better do well. Sometimes *leverage* simply means "power," as in *The scandalous emails I accidentally received from my boss gave me leverage in our next salary negotiation.*

**Liability** – A debt that one is responsible for. A negative asset, like that guy in your study group.

**Liquidation** – Firesale! Get it before it's gone!

**Liquidity** – Ability to be converted into money. Cash is liquid. Try drinking some.

**Marginal** – Marginal cost means the cost of the very last unit produced. Marginal revenue means the revenue you get from the very last unit sold. In the real world, something "marginal" may be unimportant, but marginal quantities are super-important in economics.

**Merger** – When two companies become one. Unlike an acquisition, a merger is meant to happen between "equals." That's never true; there's always a winner and a loser. Down the road, though, both sides may in fact lose, as the merged company stumbles toward destruction.

**Metric** – A measure or number by which you grade performance.

**Net Income** – What you are left with after student loan payments.

**Neoclassical** – "Standard" economics, in which people always maximize utility in rational ways and prices instantly adjust in well-functioning markets throughout the economy.

**Offline** – When your printer is acting up, or when you should really continue the conversation at a later time with a smaller group.

**Offshore Financial Center** – See *Switzerland* or *Cayman Islands*. Where you put your money once you've got oodles of it.

**Operating Income** – Operating revenue, less operating expenses. Operating income excludes income generated by "non-operations," as well as interest and tax expenses.

**Opportunity Cost** – The benefits you missed on Road B, because you took Road A instead.

**Option** – The ability to buy or sell a financial asset at a fixed price.

**Perfect Competition** – The theoretical state of affairs in a perfect market, in which many, many buyers and sellers interact to rapidly set a market-clearing price, without any one player having control over that price.

**Pitch Book** – An analysis of a deal, made by an investment bank to convince a client to commit to said deal.

**Poison Pills** – Methods implemented by a company's board of directors, seeking to prevent hostile takeovers by shareholders who are unfriendly to current management or ownership.

**Portfolio Play** – Spreading your bets, as you do when you hold a portfolio of stocks (rather than just one stock).

**Preferred Stock** – A part-stock, part-bond claim on a company. Preferred stock is more risky than regular debt but less risky than common stock.

**Prime Rate** – A reference interest rate used by banks to calculate the interest rate on certain loans, such as student loans and adjustable-rate mortgages. In the US, the prime rate is derived from an even more basic interest rate, the federal funds target rate.

**Profit** – What's left once you subtract your expenses from whatever revenues you have earned. Being in the black. Makin' money. Rollin' in the dough.

**Queue** – A line you wait in. Although common in British English, the term *queue* is rarely used by Americans outside of technical contexts.

**Resources** – The expendable folks you throw at small projects. When you yourself are referred to as a *resource,* hide.

**Salvage Value** – What you can get for your roommate's computer or the assets of a company, if you sell them off after they've lived a useful life.

**Scorecard** (Balanced Scorecard) – A tool that shows you how an operation is doing relative to a whole set of benchmarks. The point is to judge performance from several perspectives, not just one.

**Six Sigma** – A process-improvement methodology adopted by many companies. The name refers to six standard deviations away from the mean. According to the normal curve, six-sigma events happen only once in a very blue moon.

**Spot** – Current, as opposed to future. The spot market for electricity is where electricity is bought and sold for immediate delivery. A spot price is a price in the spot market.

**Supply Chain** – A system by which materials move from suppliers through various intermediaries and finally to customers.

**Sunk Cost** – Like the ante in poker, a sunk cost is what you're already in for. A common mistake is to let sunk costs affect your decisions about future courses of action.

**SWOT analysis** – A strategic evaluation of a company's Strengths, Weaknesses, Opportunities, and Threats. SWOT comes before the SWAT team but after signs of trouble.

**Synergy** – An overused word. In strategy class, *synergy* means pretty things like "working together, the whole is bigger than the sum of the parts," etc. In the real world, *synergy* is code for "who do we fire when we merge these groups?"

**Teaching Assistant (TA)** – An underqualified MBA or overqualified PhD student who leads review sections and grades the majority of papers.

**Term Sheet** – The outline of a business agreement. Often associated with private equity and venture capital deals.

**Top Line** – The revenue of a company (so called because it's on the top line of an income statement). *Marbles-R-Us is finally showing some top-line growth* means that revenues are increasing.

**Value Chain** – An end-to-end analysis of a business process, focusing on how much value is added at each stage of the process.

**Value Proposition** – Your big pitch to your customers. Why your offering is so much better than anyone else's.

**Variable Cost** – The non-fixed costs of production, such as materials and labor.

**Vertical Integration** – Owning all the links in the supply chain, starting from raw materials, all the way through distribution of finished products to end users.

**Widget** – What you call a random product when you don't care what it is, for the purposes of some academic exercise. Often made by Acme Co.

**Yield Curve** – A graphical relationship between the yield of some type of bond (its effective interest rate) and the time to maturity of the bond. This curve usually slopes upwards (longer-term bonds pay higher yields), because long-duration bonds are inherently riskier and because inflation is typically expected for the future.

## B. ACRONYM GUIDE

| Acronym | Meaning | Pronunciation |
|---|---|---|
| ABC | Always Be Closing | Pronounce each letter |
| B2B | Business-to-Business | Pronounce each letter/ number |
| B2C | Business-to-Consumer | Pronounce each letter/ number |
| BATNA | Best Alternative to Negotiated Agreement | "Bat-na" |
| CAGR | Compound Annualized Growth Rate | "Kag-err" (much like "kegger") |
| CAPM | Capital Asset Pricing Model | "Cap-M" |
| CAPX | Capital Expenditure | "Cap-X" |
| CLM | Career-Limiting Move | Pronounce each letter |
| COB | Close of Business | Pronounce each letter |
| COGS | Cost of Goods Sold | Rhymes with "hogs" |
| CRM | Consumer Resource Management | Pronounce each letter |
| CYA | Cover Your Ass | Pronounce each letter |
| DBA | Doing Business As | Pronounce each letter |
| DCF | Discounted Cash Flow | Pronounce each letter |
| DJIA | Dow Jones Industrial Average | Never spell this; call it "The Dow" or "The Dow Jones" |
| EBIT | Earnings Before Interest & Taxes | "E-bit" |

| Acronym | Meaning | Pronunciation |
|---|---|---|
| EBITDA | Earnings Before Interest, Taxes, Depreciation, & Amortization | "E-bit-da" |
| EPS | Earnings per Share | Pronounce each letter |
| FIFO | First In, First Out | "Fife-oh" |
| FOMO | Fear of Missing Out | "Foe-moh" |
| The FT | *The Financial Times* | Pronounce each letter |
| FX | Foreign Exchange | Pronounce each letter |
| FY | Fiscal Year | Pronounce as "Fiscal Year" |
| GAAP | Generally Accepted Accounting Principles | Say it like the clothing store or, as the London Tube reminds you, "Mind the ____." |
| IMF | International Monetary Fund | Pronounce each letter |
| IPO | Initial Public Offering | Pronounce each letter |
| IP | Intellectual Property or Internet Protocol | Pronounce each letter |
| IRR | Internal Rate of Return | Pronounce each letter |
| JIT | Just-in-Time | Pronounce as "Just-in-time" |
| KISS | Keep It Simple, Stupid | Say it like the heavy metal band or what you do to your sweetie |
| KPI | Key Performance Indicators | Pronounce each letter |
| LBO | Leveraged Buyout | Pronounce each letter |
| LIBOR | London Interbank Offering Rate | "Lie-bore" |
| LIFO | Last In, First Out | "Life-oh" |

| Acronym | Meaning | Pronunciation |
|---|---|---|
| LLC | Limited Liability Company | Pronounce each letter |
| LY | Last Year | Pronounce as "last year" |
| M&A | Mergers and Acquisitions | Pronounce each letter |
| NDA | Non-Disclosure Agreement | Pronounce each letter |
| NPV | Net Present Value | Pronounce each letter |
| OTC | Over the Counter | Pronounce each letter |
| PE | Private Equity | Pronounce each letter |
| P/E Ratio | Price-to-Earnings Ratio | "P.E. ratio" |
| P&L | Profit and Loss (more common than Property & Liability, as in insurance) | "P and L" |
| R&D | Research and Development | "R and D" |
| ROI | Return on Investment | Pronounce each letter |
| S&P 500 | Standard & Poor's | "S and P 500" |
| SKU | Stock Keeping Unit | "Skew," not "s-k-u" |
| VC | Venture Capital or Venture Capitalist or Variable Cost | Pronounce each letter<br>Pronounce as "variable cost" |
| WACC | Weighted Average Cost of Capital | Exactly as in "That's whack!" |
| The WSJ | *The Wall Street Journal* | "The Wall Street Journal" |
| YOY | Year Over Year | Pronounce as "year over year" |

## ACKNOWLEDGMENTS

We had incredible help and support in writing this book and its predecessor, *Case Studies & Cocktails*, which covers not only the academic side of business school but the social side, the job-hunting side, even the side *before* you get to school. For that book, we talked to nearly two hundred business school students (including many former GMAT students of ours) and administrators. We are, again, profoundly grateful for their insights and perspectives.

The Kaplan Publishing team roundly deserves our applause. May that sound also echo out to all the Manhattan Prep folks who made *Case Studies & Cocktails* possible in the first place.

***Carrie:*** It's hard to believe all that's transpired since *Case Studies & Cocktails* first appeared, but I couldn't be more grateful for that experience and for the opportunity to revisit it now. Thanks again to all who made that endeavor possible, and to my family and friends, who continue to provide the a solid grounding to any whims of irrational exuberance. And to my partner, Minsok, who creates unmeasurable value in my life.

***Chris:*** I'd like to send a shout-out to everyone who got name-checked in *Case Studies & Cocktails*. For the sake of space, I'll just explicitly thank Mom and Dad and my siblings one more time. And now that my wife, Kathryn, is mother to our two young wonderfuls, Seamus and Eleanor, born since the previous book, all three deserve a new dose of gratitude. Thanks for teaching, and often schooling, me every day.

## ABOUT THE AUTHORS

Carrie Shuchart holds an A.B. in social studies from Harvard University and an MBA from Columbia Business School. She has been an editorial analyst at *The Atlantic,* a freelance television producer, and an entrepreneur. While at Columbia, she focused her efforts on *Follies,* the school's semiannual comedy show, for which she acted, directed, and made some pretty poor attempts at dancing.

She began teaching with Manhattan Prep (then Manhattan GMAT) in 2008 as a way to fund her startup habit and her shoe collection. In 2010, Carrie joined the Los Angeles office of McKinsey & Co. as a consultant. She now serves as Director, Head of Communications for McKinsey New Ventures. She lives in Minneapolis, Minnesota, with her partner.

Chris Ryan has an A.B. in physics from Harvard University and an MBA from the Fuqua School of Business at Duke University. He taught high school science for several years, at first through Teach for America. At Fuqua, Chris was head TA of the core statistics and finance courses, as well as Curriculum Representative and *Fuquavision* copresident.

After a stint at McKinsey, Chris joined Manhattan Prep in 2003 and helped it grow into the largest GMAT prep provider in the world. He is now the Executive Director of Product Strategy for the Admissions Group at Kaplan Test Prep, which includes Manhattan Prep. He lives in Carrboro, North Carolina, with his wife and two children.